OFFICIAL SQA PAST PAPERS WITH ANSWERS

STANDARD GRADE | CREDIT

BIOLOGY
2007-2011

BrightRED
PUBLISHING

Publisher's Note

We are delighted to bring you the 2011 Past Papers and you will see that we have changed the format from previous editions. As part of our environmental awareness strategy, we have attempted to make these new editions as sustainable as possible.

To do this, we have printed on white paper and bound the answer sections into the book. This not only allows us to use significantly less paper but we are also, for the first time, able to source all the materials from sustainable sources.

We hope you like the new editions and by purchasing this product, you are not only supporting an independent Scottish publishing company but you are also, in the International Year of Forests, not contributing to the destruction of the world's forests.

Thank you for your support and please see the following websites for more information to support the above statement –

www.fsc-uk.org

www.loveforests.com

© Scottish Qualifications Authority

All rights reserved. Copying prohibited. No part of this publication may be reproduced, stored in a retrieval system, or transmitted in any form or by any means, electronic, mechanical, photocopying, recording or otherwise.

First exam published in 2007.
Published by Bright Red Publishing Ltd, 6 Stafford Street, Edinburgh EH3 7AU
tel: 0131 220 5804 fax: 0131 220 6710 info@brightredpublishing.co.uk www.brightredpublishing.co.uk

ISBN 978-1-84948-159-5

A CIP Catalogue record for this book is available from the British Library.

Bright Red Publishing is grateful to the copyright holders, as credited on the final page of the Question Section, for permission to use their material. Every effort has been made to trace the copyright holders and to obtain their permission for the use of copyright material. Bright Red Publishing will be happy to receive information allowing us to rectify any error or omission in future editions.

STANDARD GRADE | CREDIT

2007

[BLANK PAGE]

C

FOR OFFICIAL USE

	KU	PS

Total Marks

0300/402

NATIONAL
QUALIFICATIONS
2007

MONDAY, 21 MAY
10.50 AM – 12.20 PM

BIOLOGY
STANDARD GRADE
Credit Level

Fill in these boxes and read what is printed below.

Full name of centre

Town

Forename(s)

Surname

Date of birth
Day Month Year

Scottish candidate number

Number of seat

1 All questions should be attempted.

2 The questions may be answered in any order but all answers are to be written in the spaces provided in this answer book, and must be written clearly and legibly in ink.

3 Rough work, if any should be necessary, as well as the fair copy, is to be written in this book. Additional spaces for answers and for rough work will be found at the end of the book. Rough work should be scored through when the fair copy has been written.

4 Before leaving the examination room you must give this book to the invigilator. If you do not, you may lose all the marks for this paper.

SCOTTISH
QUALIFICATIONS
AUTHORITY

©

Marks | KU | PS

1. The graph shows the changes in the population of bacteria in a fermenter.

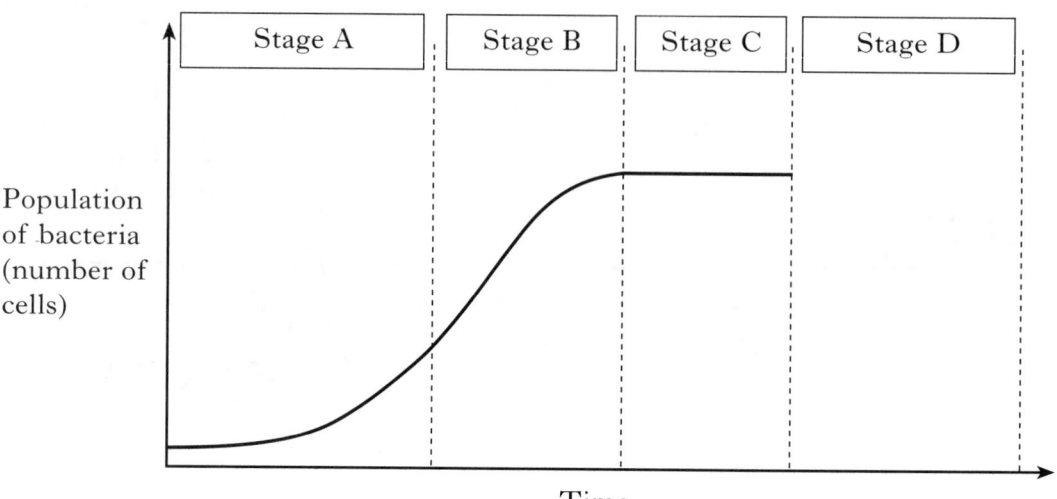

(a) (i) Describe the changes in population of the bacteria during Stage B.

_____

_____   **1**

(ii) Give a reason for the changes in population shown during Stage B on the graph.

_____

_____   **1**

(iii) Complete Stage D on the graph to show the effect of an increasing death rate on the population of bacteria.   **1**

(b) Some bacteria can be grown on industrial waste materials to provide valuable products, such as animal foodstuffs.

State **one** way in which the nutritional value of the product has been increased.

_____   **1**

2. The diagram shows some of the stages in the nitrogen cycle.

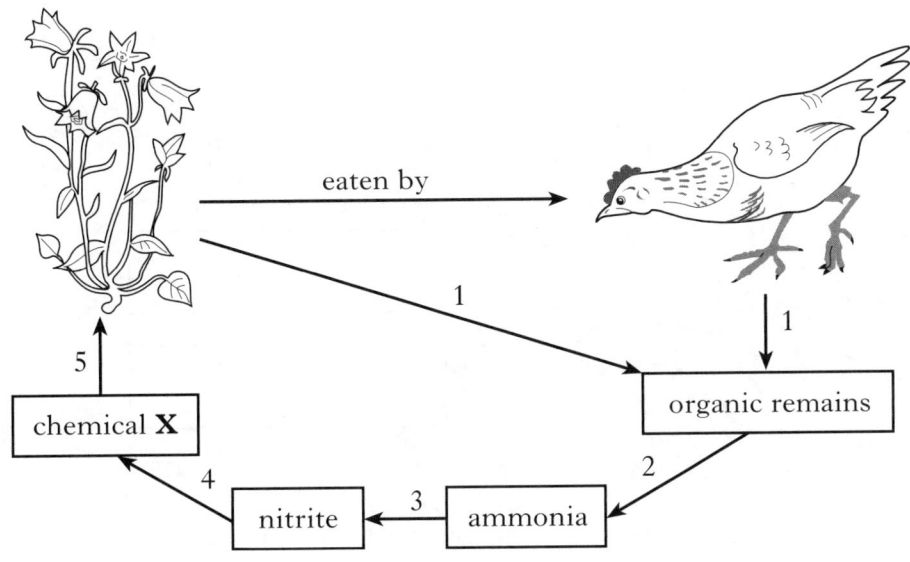

(a) Complete the table by giving a number from the diagram to match each of the named stages.

Stage	Number
Absorption	
Death	
Nitrification	
Decomposition	

2

(b) Name chemical **X**.

_____

1

(c) Name the type of organism responsible for Stage 3.

_____

1

[Turn over

Marks | KU | PS

3. (*a*) Carbon dioxide is used during photosynthesis to produce sugar.

(i) Complete the diagram below to show the fates of carbon dioxide after photosynthesis has taken place.

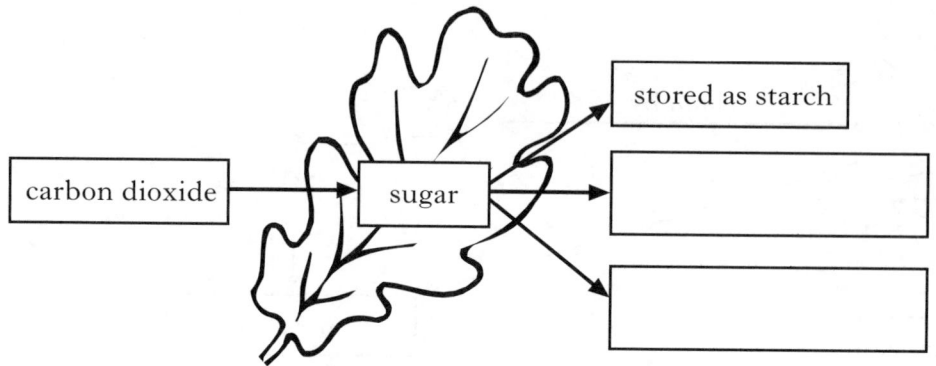

stored as starch

carbon dioxide → sugar

2

(ii) Explain why soot deposits on leaves reduce the rate of photosynthesis.

_____

_____

1

(*b*) (i) Draw an **X** on the following diagram to show where the pollen tube reaches when its growth is completed.

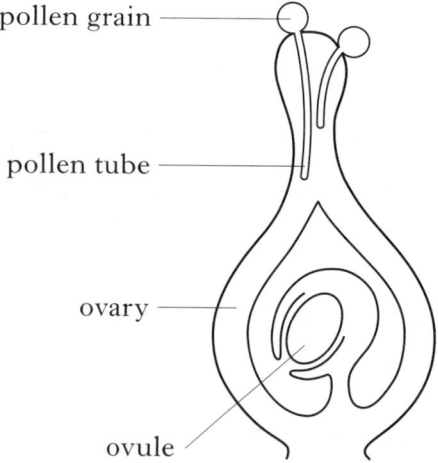

pollen grain

pollen tube

ovary

ovule

1

(ii) Describe the function of the pollen tube.

_____

_____

1

Marks | KU | PS

3. (continued)

(c) Tropical rain forests are being destroyed to clear land for farming. This leads to a reduction in the number of plant species.

Explain why this might lead to the extinction of some animal species.

_____

1

(d) The diagrams show features of some newly discovered plants.

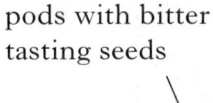

scented flowers with brightly coloured petals

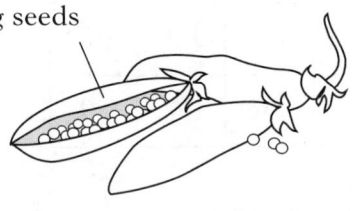

pods with bitter tasting seeds

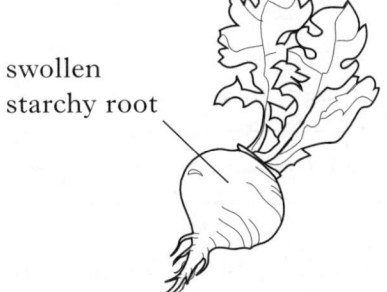

tough stem with strong fibres

swollen starchy root

Select **one** of the plant features and describe a likely use for it.

Plant feature _____

Likely use _____

_____

1

[Turn over

Marks | KU | PS

4. The following investigation was set up to examine the effects of stirring on the digestion of protein.

Each piece of protein was weighed every two hours.

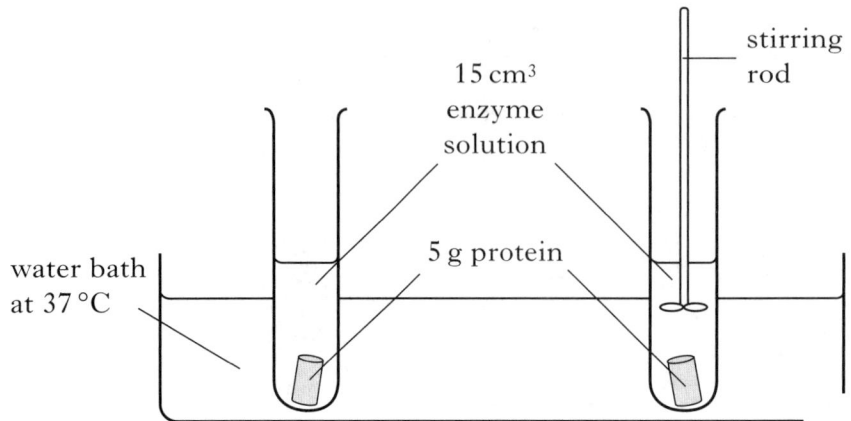

The results are shown in the table.

		Time (hours)					
		0	2	4	6	8	10
Mass of protein (g)	not stirred	5·0	4·7	4·3	3·8	3·2	2·5
	stirred	5·0	4·4	3·6	2·6	1·4	0·0

(a) Use the data in the table to complete the line graph below.

(An additional graph, if needed, will be found on page 25.)

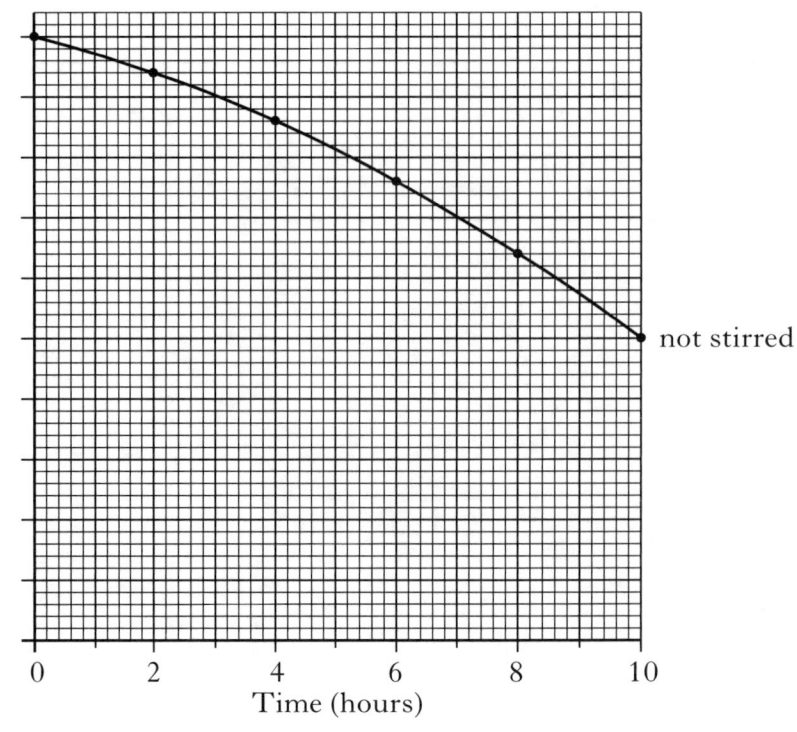

2

Marks | KU | PS

4. **(continued)**

(*b*) Which type of enzyme would produce the results shown?

_____

1

(*c*) When the protein was completely digested, no solid material remained in the tube. Explain why.

_____

1

(*d*) Name **one** factor, not already mentioned, which would need to be the same in each tube at the start of the investigation.

_____

1

(*e*) Suggest how the investigation could be improved to provide a more reliable measurement of the difference which stirring made.

_____

1

(*f*) Stirring increased the rate at which the protein was digested. Explain why this happened.

_____

1

(*g*) In the body, the stomach achieves a similar effect to stirring. Describe how this happens.

_____

1

[Turn over

5. The diagram represents a microscopic part of a kidney.

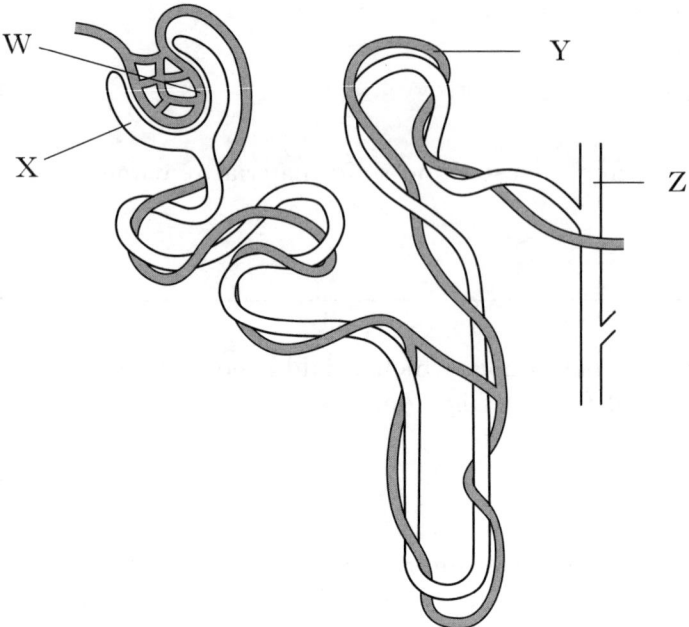

(*a*) Complete the table to show the names and functions of the structures shown on the diagram.

	Name	*Function*
W	glomerulus	
X		collection of filtrate
Y		reabsorption
Z	collecting duct	

2

Marks | KU | PS

5. (continued)

(b)   The table shows information about kidney function.

Fluid	Component (g per 100cm³)				
	urea	glucose	amino acids	salts	proteins
blood plasma	0·03	0·10	0·05	0·9	8·0
glomerular filtrate	0·03	0·10	0·05	0·9	none
urine	1·75	none	none	0·90–3·60	none

(i)   In which organ is urea produced and how is it transported to the kidneys?

Organ _____

Means of transport _____   **1**

(ii)   Name **one** component in the table which can pass through the wall of the glomerulus, and **one** component which cannot.

Can pass through _____

Cannot pass through _____   **1**

(c)   In one investigation, the kidneys of an adult male were found to filter 1254 cm³ of blood per minute. This produced 114 cm³ of filtrate per minute and 1·2 cm³ of urine per minute.

(i)   Express these volumes as a simple whole number ratio.

Space for calculation

_____ : _____ : _____
blood   filtrate   urine   **1**

(ii)   Using the results of this investigation and information from the table, calculate the mass of urea which would be excreted by this person in 24 hours.

Space for calculation

_____ g   **1**

Marks | KU | PS

6. The brown shrimp is found all round our coastline.

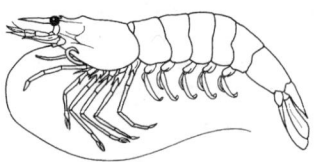

The graph shows shrimp activity and changes in their environment over a 48 hour period.

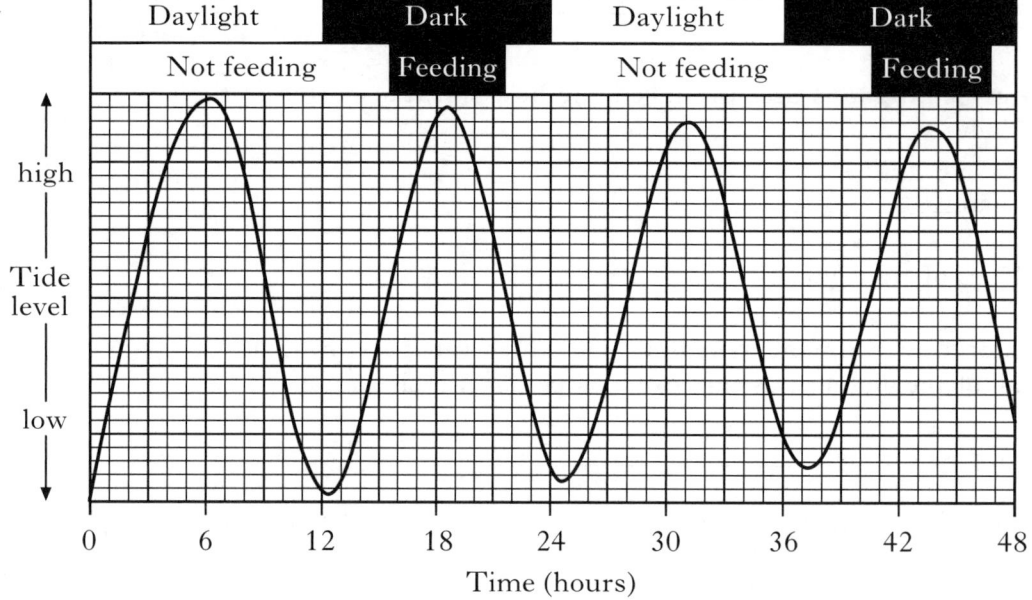

(a) How many high tides occurred during the two days shown?

_____

1

(b) Describe the conditions necessary for the shrimps to feed.

_____

_____

2

(c) Explain the significance of the behaviour shown to the survival of the shrimps.

_____

_____

1

Marks | KU | PS

7. A flower petal was examined under the microscope and then placed in a concentrated salt solution for 30 minutes. It was then re-examined under the microscope.

The diagrams show a cell from the petal before and after being in the solution.

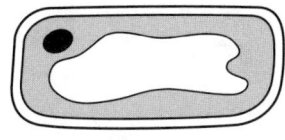

before

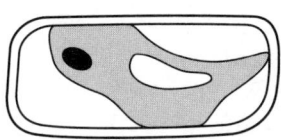

after

(a) (i) The movement of water caused the change in the appearance of the cell. What name is given to this movement of water?

_____

1

(ii) In terms of water concentration, explain **why** this movement of water took place.

_____

_____

1

(b) Name **one** substance, other than water, which must be able to pass into a cell for its survival.

_____

1

(c) The diagram below shows a group of cells as seen under a microscope. The field of view was 2 mm in diameter.

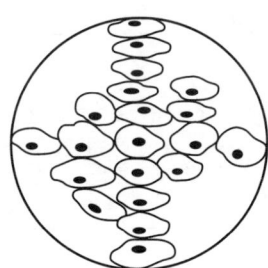

Calculate the average length and width of the cells.

Space for calculation

Average length _____ mm

Average width _____ mm

1

[Turn over

Marks | KU | PS

8. (a) The diagram shows a method used to investigate the energy content of a variety of foods.

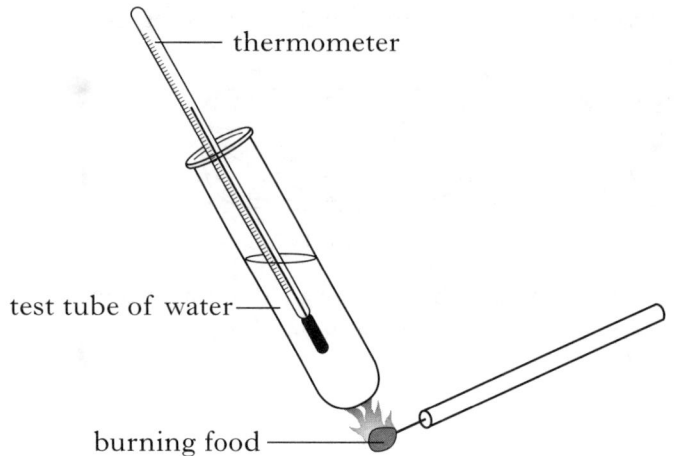

- thermometer

test tube of water

burning food

The rise in temperature can be used to calculate the energy content of each food in kilojoules.

The results are shown in the table.

Type of food	mass (g)	energy content (kilojoules)
cheese	1·0	17·0
fish	1·0	0·5
steak	1·0	13·9
carrot	1·0	1·8
apple	1·0	2·5

(i) State **two** factors, not already mentioned, that should be kept constant for a valid comparison to be made between the foods.

1 _____

2 _____   2

(ii) Suggest why the energy contents found in the investigation might not have been as high as expected.

_____

_____   1

8. (a) (continued)

(iii) The energy content of each food was calculated using the following formula.

Energy content (kilojoules) = temperature rise $\times 0.21$

Calculate the energy content of 1g of chicken, if it raised the temperature of the water by 30 °C.

Space for calculation

_____ kilojoules per gram   1

(b) Give **one** reason, other than providing heat, why cells need energy from food.

_____   1

(c) Which component of food provides most energy per gram?

_____   1

[Turn over

9. The diagram below shows a cross-section through a joint.

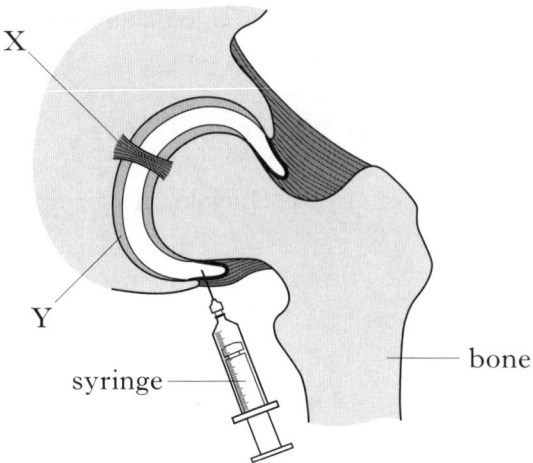

(a) Name and describe the functions of parts X and Y on the diagram.

Part X   Name _____

Function _____

_____   **1**

Part Y   Name _____

Function _____

_____   **1**

(b) Some of the synovial fluid from inside a joint can be removed for medical tests using a syringe as shown in the diagram.

   (i) Name the part of the joint which produces the synovial fluid and describe the function of the fluid.

        Produced by _____

        Function _____   **1**

Marks | KU | PS

9. (*b*) (continued)

(ii) The table below describes the features of the fluid which lead to the diagnosis of several joint abnormalities.

		Feature of synovial fluid		
		Viscosity	*Cloudiness*	*Colour*
Diagnosis	*Normal*	high	zero	light yellow
	Inflammation	low	slight	dark yellow
	Infection	low	high	dark yellow
	Blood leakage	intermediate	high	pink

Use the information from the table to complete the paired statement key to identify the diagnoses.

1. Fluid pink .... Blood leakage

   Fluid not pink .... go to 2

2. Low viscosity ....

   High viscosity ....

3. [_____] .... Infection

   [_____] .... [_____]

2

[Turn over

Marks | KU | PS

10. Read the following passage and answer the questions based on it.

Invasion of the Chinese Mitten Crab
Adapted from *Biological Sciences Review*, Volume 15, Number 2.

The Chinese mitten crab, *Eriocheir sinensis*, lives in fresh water as an adult, but it breeds in the lower reaches of estuaries and spends part of its early life in seawater.

It looks different from other crabs. Its claws are covered in a coating of fine brown hairs resembling mittens. This type of crab is a problem because it burrows into river banks, causing them to collapse and silt up river channels.

The mitten crab is not native to Europe. They were recorded in the River Thames in the 1930s. Their larvae may have been transported to the river in ships' ballast water and released during dumping of this water before the ship took on cargo.

Adult mitten crabs have been known to travel thousands of kilometres in freshwater at up to 18 km per day. The young crabs, when migrating upriver, seem to be mainly herbivorous. As they grow, they become omnivorous, eating vegetation, crustaceans, insects and dead fish—in fact anything they can get a hold of! Not only is this a problem for the plants and animals that they are eating, but also they compete with native species, such as freshwater crayfish, for food.

(*a*) How does the appearance of the Chinese mitten crab differ from other crabs?

_____   1

(*b*) State the type of environment the Chinese mitten crabs are found in at each of the following stages in their life.

(i) Early years_____

(ii) Breeding times _____

(iii) Mature adults_____   1

(*c*) How is it thought that the Chinese mitten crabs arrived in Britain?

_____   1

DO NOT WRITE IN THIS MARGIN

Marks | KU | PS

10. (continued)

(*d*) Describe **one** problem the Chinese mitten crab causes to the habitat and **one** problem it causes to the native community.

Habitat ________________________________ 1

Community ________________________________ 1

(*e*) Describe the changes in its diet as a young adult mitten crab grows.

____

________________________________ 1

(*f*) When moving at their maximum speed, how long would it take an adult mitten crab to travel the whole length of a 45 km river?

Space for calculation

__________ days 1

[Turn over

11. (a) Lactic acid is a waste product from one type of respiration. What type of respiration produces lactic acid?

_____

1

(b) The lactic acid content of the blood of a professional cyclist was measured while cycling at different speeds.

The graph shows the results of these measurements taken at the start of the racing season and at the end.

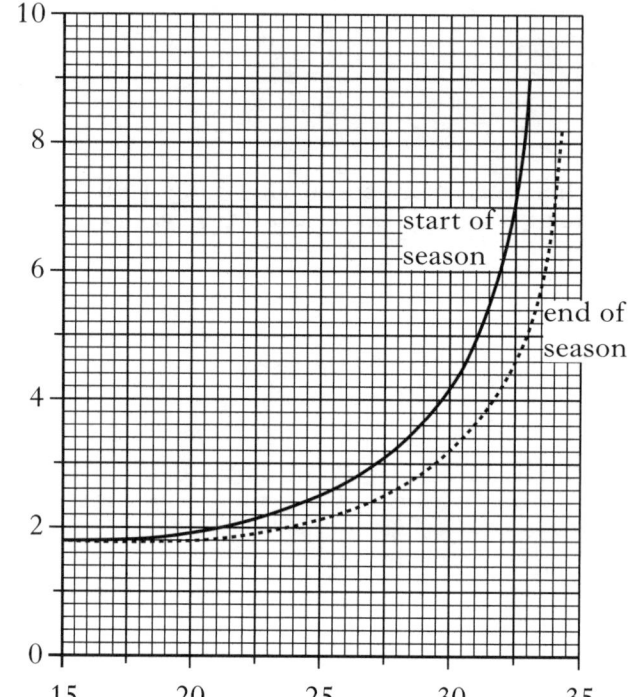

Lactic acid concentration (nM)

speed (miles per hour)

(i) What was the lactic acid concentration when the cyclist was travelling at 15 miles per hour?

_____ nM

1

(ii) At the start of the season, what was the speed of the cyclist when he was producing 50% of his maximum lactic acid concentration?

_____ miles per hour

1

Marks | KU | PS

11. **(b)** **(continued)**

(iii) When lactic acid concentration rises above 2·5 nM, the leg muscles quickly lose power and become painful.

1 What name is given to this condition?

_____

1

2 What is the maximum speed this cyclist could maintain at the start of the season?

_____ miles per hour

1

(iv) The graph shows that training improves the efficiency of muscles. Other than muscle, name **two** organs whose efficiency is improved by training.

1 _____

2 _____

1

[Turn over

12. Tongue-rolling is an inherited characteristic. The diagram below shows the pattern of its inheritance in one family.

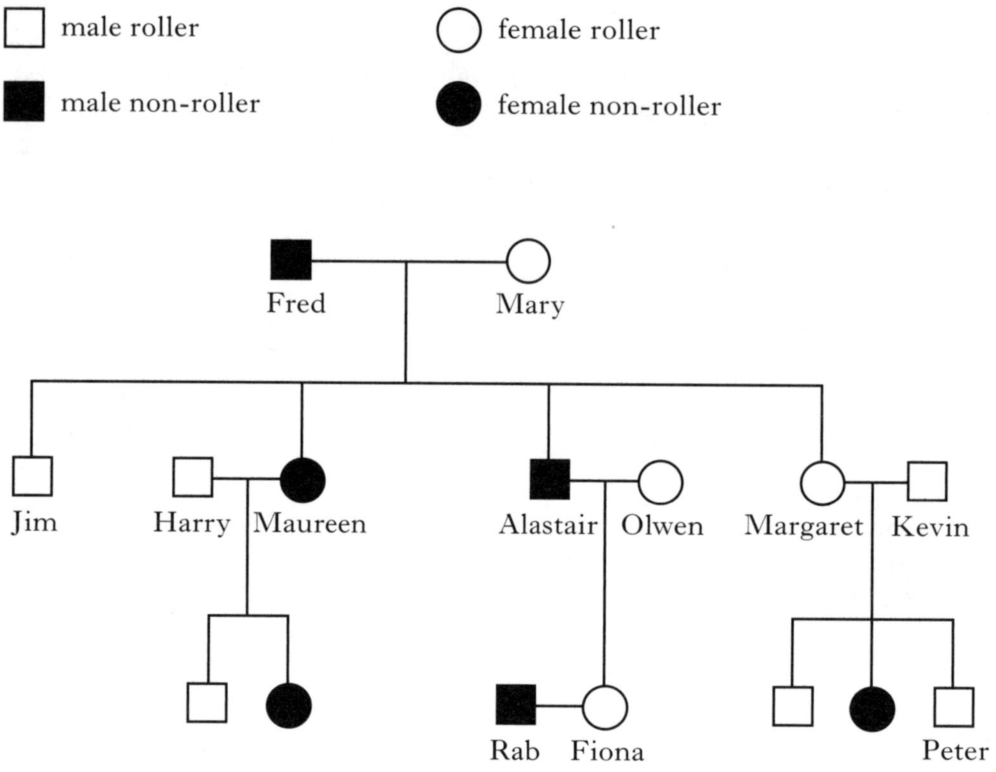

☐ male roller   ○ female roller

■ male non-roller   ● female non-roller

(a) (i) Using **R** for the dominant form of the gene and **r** for the recessive form, state the genotypes of:

1 Maureen _____

2 Jim   _____

3 Kevin   _____

(ii) If Rab and Fiona have a child, what are the chances of the child being able to roll its tongue?

Space for working

_____

(iii) Which of the original parents could be described as true-breeding?

Tick (✓) the correct box.

Fred ☐   Mary ☐

Both ☐   Neither ☐

(iv) Name a tongue-roller from the F₁ generation.

_____

Marks: 2, 1, 1, 1

DO NOT WRITE IN THIS MARGIN

12. (continued) *Marks* KU PS

(*b*) Explain why the proportions of the offspring phenotypes from genetic crosses are not always exactly as predicted.

____

____ **1**

(*c*) What term is used for the different forms of the same gene?

____ **1**

[Turn over

13. The diagram shows an industrial fermenter. It is fitted with a number of taps which allow substances to be added or removed.

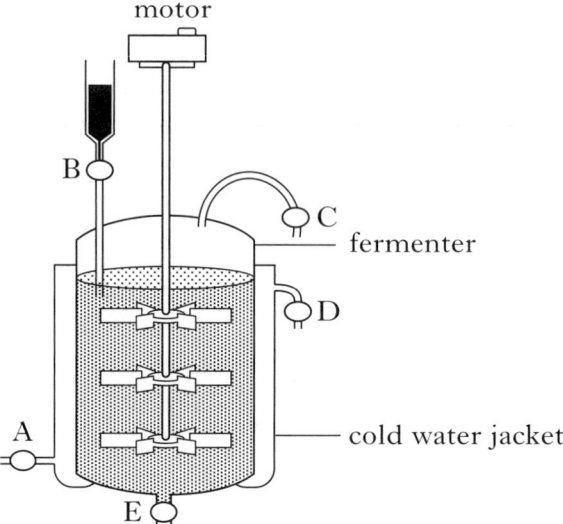

(a) Which of the taps, A, B, C, D or E, would open to

(i)   add nutrients to the mixture? _____

(ii) remove waste gases?   _____

(iii) drain off the products?   _____   **2**

(b) The fermenter should be kept at 35 °C. Explain why the water jacket around the fermenter should be cold.

_____

_____   **1**

(c) After fermentation is complete, the fermenter is drained and the useful product is separated. New starting ingredients can then be added to the fermenter.

(i)   What name is given to this type of process?

_____   **1**

(ii) When the vessel is empty, it is treated to destroy residual spores of fungi and bacteria. How could this be done?

_____

_____   **1**

Marks | KU | PS

13. (continued)

(d) Barley malt extract, water and yeast were placed in the fermenter and left for several days.

The rate of fermentation was measured and the results are shown in the graph below.

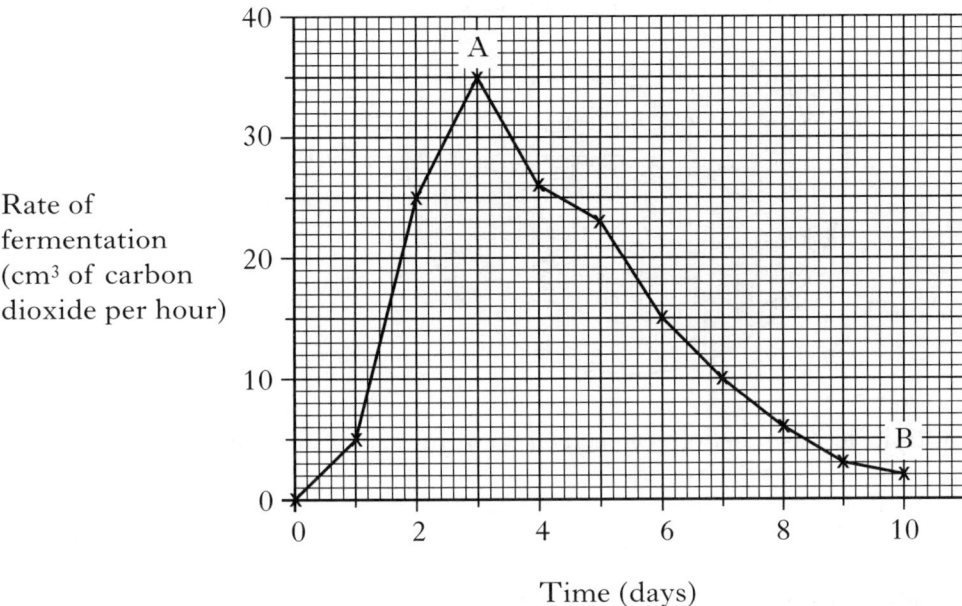

Rate of fermentation (cm³ of carbon dioxide per hour)

Time (days)

(i) Describe the changes in the rate of fermentation over the ten days.

_____

_____ 2

(ii) Suggest a reason for the change in the rate of fermentation between points A and B.

_____ 1

(iii) Why must the barley be malted before it can be used by the yeast?

_____

_____ 1

[Turn over for Question 14 on *Page twenty-four*

Marks | KU | PS

14. A nutrient agar plate was covered evenly with a suspension of bacteria. A multidisc was placed on the surface of the agar. Each of the six ends of the multidisc contained a different antibiotic.

The diagram shows the agar plate after incubation.

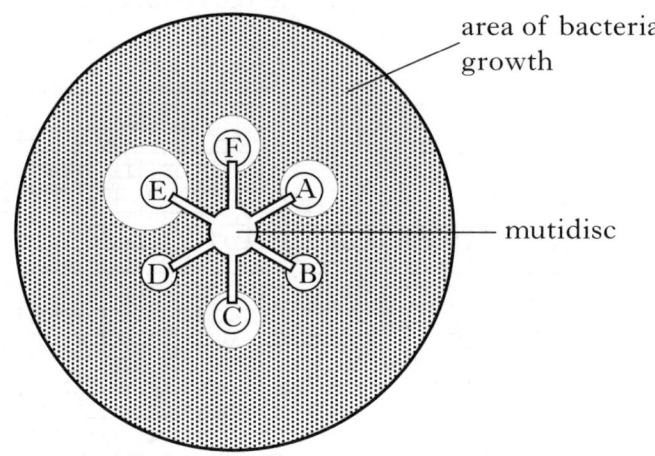

area of bacterial growth

mutidisc

(a) Which antibiotic was most effective at preventing bacterial growth?

_____

1

(b) To which antibiotics were the bacteria resistant?

_____

1

(c) Explain the need for a range of antibiotics in the treatment of diseases caused by bacteria.

_____

_____

1

[END OF QUESTION PAPER]

ADDITIONAL GRAPH PAPER FOR QUESTION 4(a)

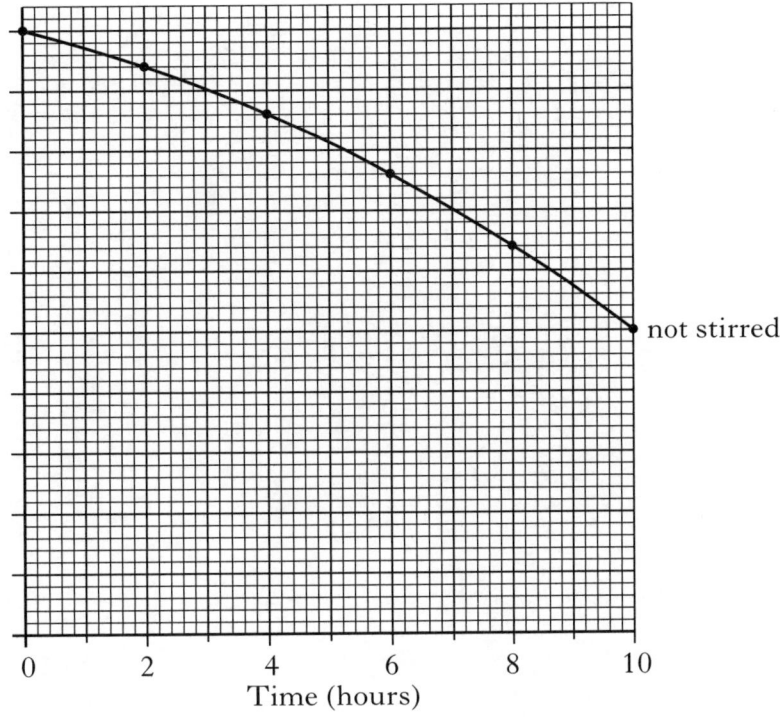

Time (hours)

[Turn over

SPACE FOR ANSWERS
AND FOR ROUGH WORKING

STANDARD GRADE | CREDIT

2008

[BLANK PAGE]

FOR OFFICIAL USE

C

KU	PS

Total Marks

0300/402

NATIONAL
QUALIFICATIONS
2008

TUESDAY, 27 MAY
10.50 AM – 12.20 PM

BIOLOGY
STANDARD GRADE
Credit Level

Fill in these boxes and read what is printed below.

Full name of centre

Town

Forename(s)

Surname

Date of birth

Day Month Year   Scottish candidate number   Number of seat

1 All questions should be attempted.

2 The questions may be answered in any order but all answers are to be written in the spaces provided in this answer book, and must be written clearly and legibly in ink.

3 Rough work, if any should be necessary, as well as the fair copy, is to be written in this book. Additional spaces for answers and for rough work will be found at the end of the book. Rough work should be scored through when the fair copy has been written.

4 Before leaving the examination room you must give this book to the invigilator. If you do not, you may lose all the marks for this paper.

Marks | KU | PS

1. (a) A comparison was made between the types of invertebrate animals living on the branches and leaves on an oak tree with those living on a beech tree.

Samples were collected as shown below.

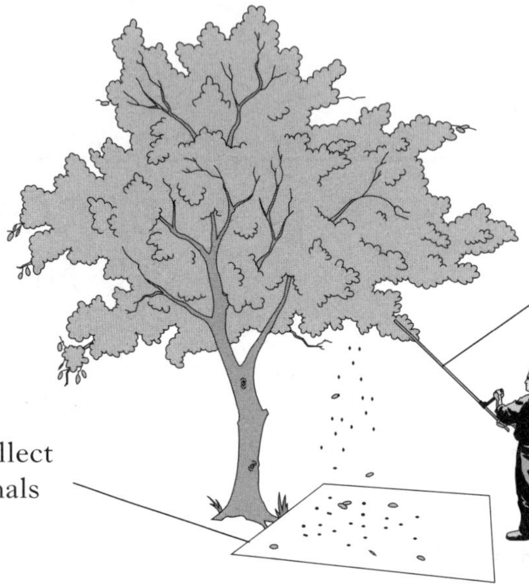

stick to shake branches

sheet to collect fallen animals

(i) Give **two** variables which should be kept constant to make the comparison valid when using this technique.

1 _____

2 _____   **1**

(ii) The samples collected were not representative of all the invertebrates living on the trees. Suggest a reason for this.

_____   **1**

(iii) Measurement of abiotic factors such as light intensity may be recorded at the same time as sampling. Identify a possible source of error for a **named** measurement technique and explain how it might be minimised.

Measurement technique _____

Source of error _____

How to minimise it _____

_____   **1**

1. (continued)

(*b*) An investigation was carried out into the effect of light intensity on the distribution of a plant species. At eight different measurement points in a garden, the average light intensity was measured and the percentage ground cover of the plant was recorded.

The results are shown below.

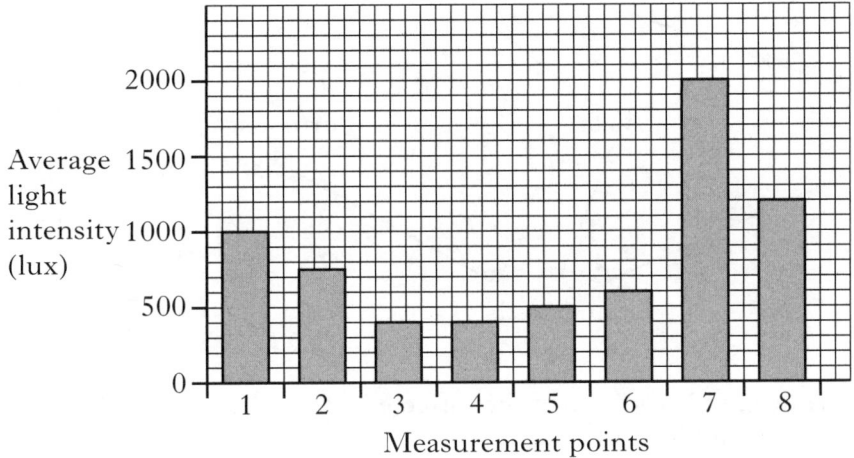

Measurement points	Ground cover of the plant (%)
1	85
2	65
3	20
4	20
5	30
6	35
7	100
8	90

(i) 1   What is the light intensity in the garden where the ground cover of the plant was 100%?

_____ lux

Marks KU PS

1

2   What was the percentage ground cover of the plant when the light intensity was 750 lux?

_____ %

1

(ii) What is the relationship between light intensity and percentage ground cover of the plant?

_____

_____

1

(*c*) Explain how light intensity affects the distribution of the plants in the garden.

_____

_____

1

Marks | KU | PS

2. (*a*) The diagram shows part of a food web from a forest.

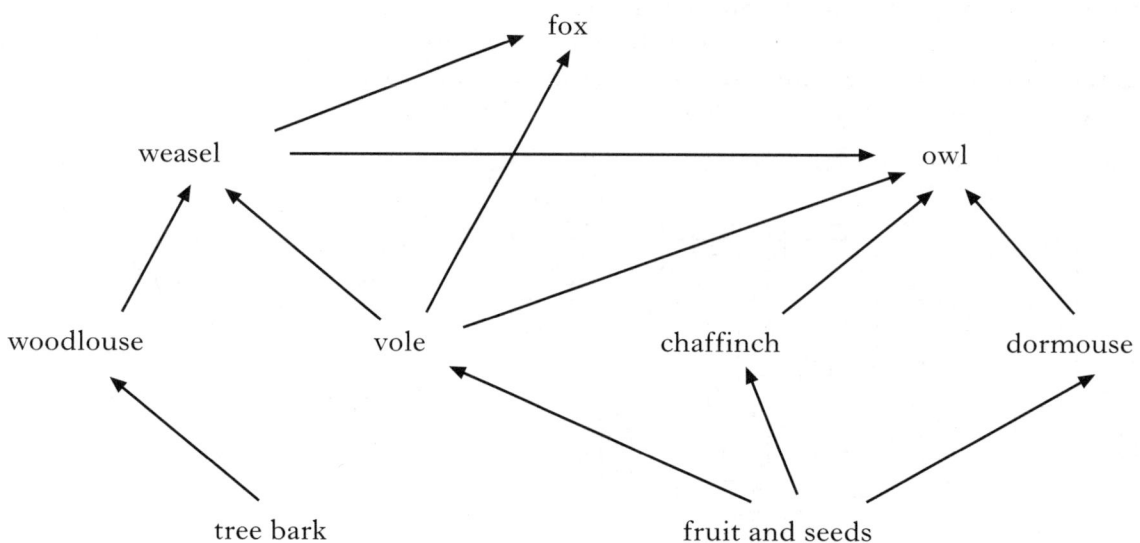

(i) The numbers of dormice and owls may be affected if the chaffinches were removed from the food web.

1 Underline **one** answer in the brackets and give an explanation.

The dormouse population would $\left\{ \begin{array}{l} \text{increase} \\ \text{decrease} \\ \text{stay the same} \end{array} \right\}$.

Explanation _____

_____   1

2 Underline **one** answer in the brackets and give an explanation.

The owl population would $\left\{ \begin{array}{l} \text{increase} \\ \text{decrease} \\ \text{stay the same} \end{array} \right\}$.

Explanation _____

_____   1

(ii) Select a food chain from the web which is made up of four stages.

_____ → _____ → _____ → _____   1

DO NOT WRITE IN THIS MARGIN

Marks KU PS

2. (continued)

(*b*) A food chain from the ocean is shown below.

plankton → krill → blue whale

Which population in the food chain has the smallest biomass?

______________________________ **1**

[Turn over

Marks | KU | PS

3. (*a*) The grid contains the names of some components of food.

carbon	A	hydrogen	B	amino acids	C
nitrogen	D	simple sugar	E	glycerol	F
fatty acids	G	oxygen	H	water	I

Use letters from the grid to identify the following:

(i) The sub-units of protein molecules _____   1

(ii) The sub-units of fat molecules _____ and _____   1

(iii) An element found in protein but not in starch _____   1

(*b*) Name the structures in the small intestine which provide an increased surface area for absorption.

_____   1

(*c*) Urea is produced in the liver from the breakdown of digested food molecules. From which food molecules is urea produced?

_____   1

Marks | KU | PS

4. (a) The diagram shows part of the human breathing system.

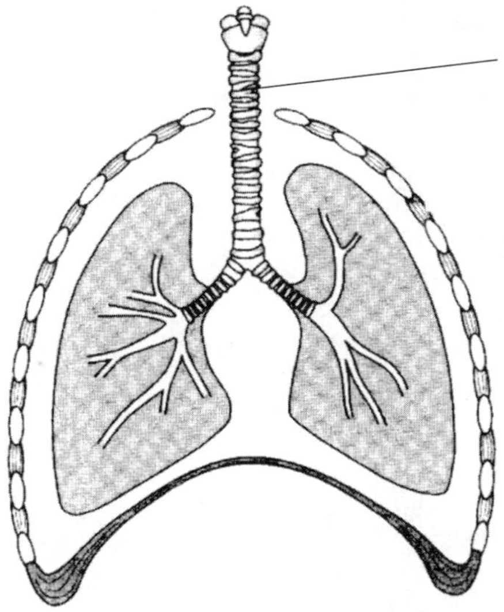

cartilage rings

Describe the function of the cartilage rings.

_____   1

(b) (i) Name the sticky substance that traps inhaled dust particles.

_____   1

(ii) Explain how the trapped particles are removed from the breathing system.

_____

_____   1

(c) As blood passes through capillary networks in the lungs, oxygen and carbon dioxide are exchanged between the blood and the air sacs.

(i) Describe **one** feature of a capillary network which allows efficient gas exchange.

_____

_____   1

(ii) Name the structures in blood that contain haemoglobin.

_____   1

(iii) Explain the function of haemoglobin in the transport of oxygen.

_____   1

Marks | KU | PS

5. (*a*) The diagram represents phloem tissue from the stem of a plant.

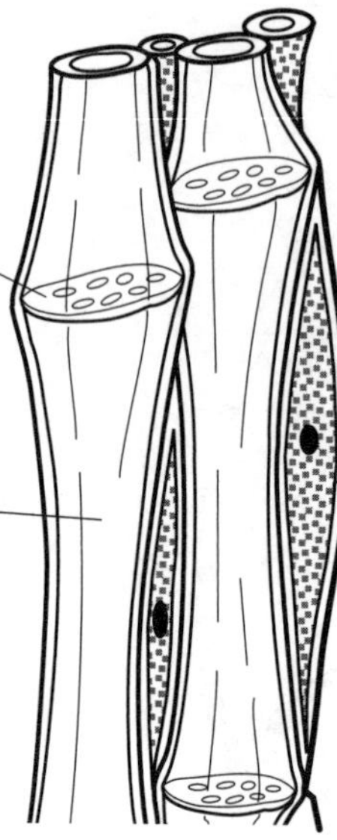

Structure A

Sieve tube

Cell B

(i) Name Structure A and Cell B.

Structure A _____

Cell B _____

2

(ii) State the function of phloem.

_____

1

(*b*) (i) Name the leaf tissue where stomata are found.

_____

1

(ii) Name the cells which control the opening and closing of stomata.

_____

1

Marks | KU | PS

5. (continued)

(c) Leaves were placed in tubes as shown below.

The tubes were left in bright light.

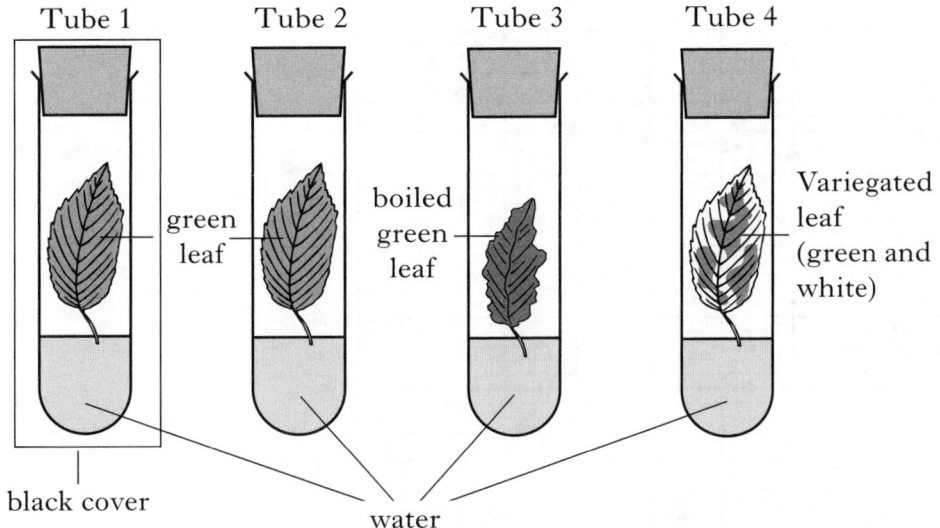

Tube 1   Tube 2   Tube 3   Tube 4

green leaf

boiled green leaf

green leaf

Variegated leaf (green and white)

black cover   water

For each of the tubes, tick (✓) the appropriate box in the table to indicate which processes will take place in the leaves.

Process / Tube	Only photosynthesis	Only respiration	Both	Neither
1				
2				
3				
4				

2

[Turn over

6. (*a*) The graph shows the number of kidney transplants carried out and the number of patients waiting for a transplant in the UK between 1996 and 2005.

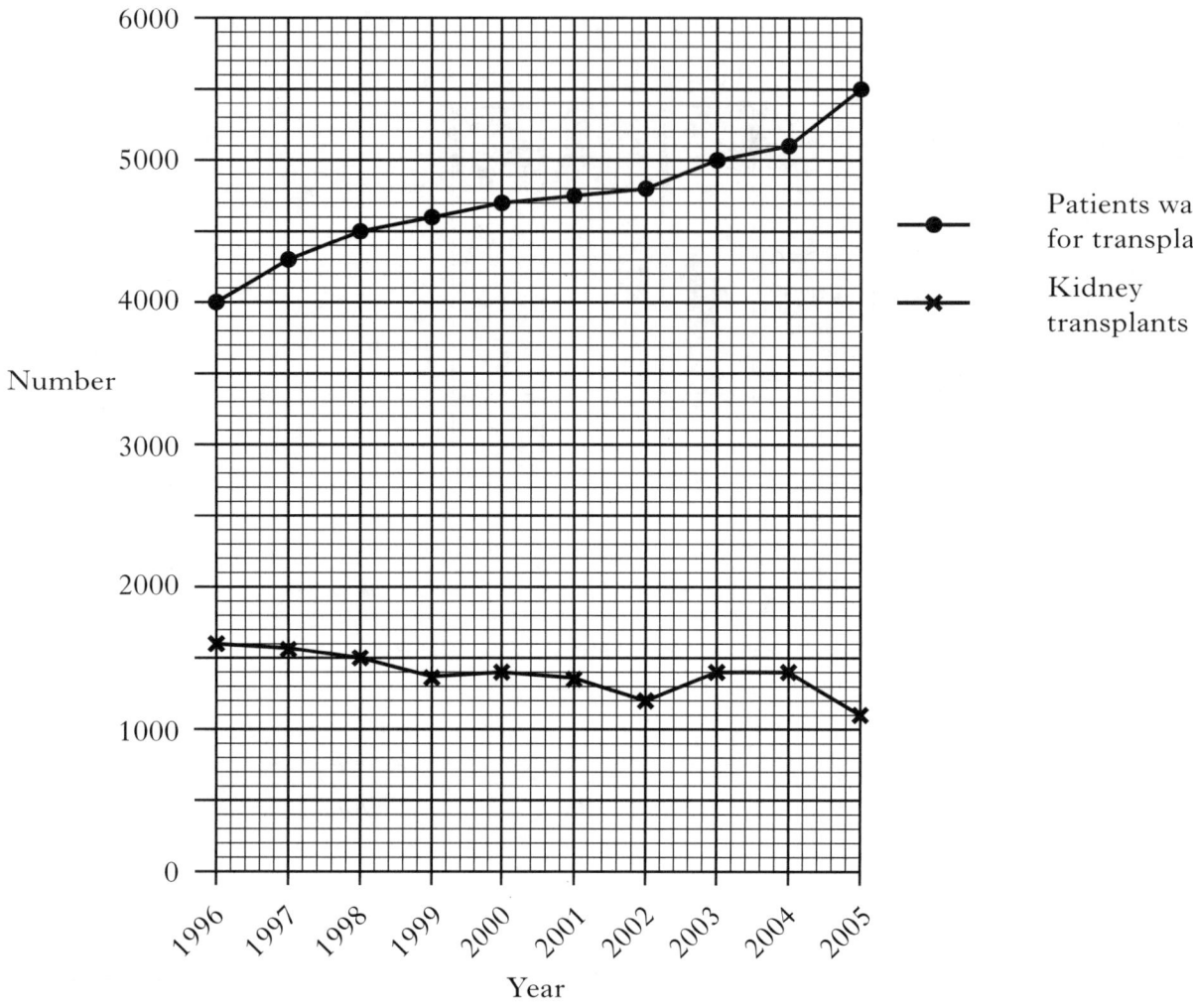

Patients waiting for transplant

Kidney transplants

(i) Calculate the average yearly increase in the number of patients waiting for a transplant from 2000 to 2005.

Space for calculation.

Average yearly increase _____ patients per year.

1

DO NOT WRITE IN THIS MARGIN

6. (a) (continued)

(ii) Calculate the simple whole number ratios of patients waiting for a transplant to the number of kidney transplants carried out for 1996 and for 2005.

Space for calculation.

1996 _____ : _____

2005 _____ : _____

         patients waiting   transplants
         for a transplant   carried out

Marks: 1

(iii) The following statements refer to the data in the graph.

Tick (✔) the box(es) of the correct statement(s).

The number of patients waiting for a transplant increased every year. ☐

The number of transplants carried out decreased every year. ☐

The difference between the number of patients waiting for a transplant and the number of transplants carried out increased every year. ☐

Marks: 1

(b) Give **one** advantage and **one** disadvantage of treating kidney failure by transplant compared to treatment using a dialysis (kidney) machine.

Advantage _____

_____

Marks: 1

Disadvantage _____

_____

Marks: 1

[Turn over

Marks | KU | PS

7. An investigation was carried out into the effect of the mineral boron on the growth of young trout.

Immediately after fertilisation, trout eggs were placed in distilled water containing different concentrates of boron.

After hatching, young trout survive on food from their yolk sac for a maximum of four weeks. The graph below shows the average lengths of the young trout three weeks after hatching.

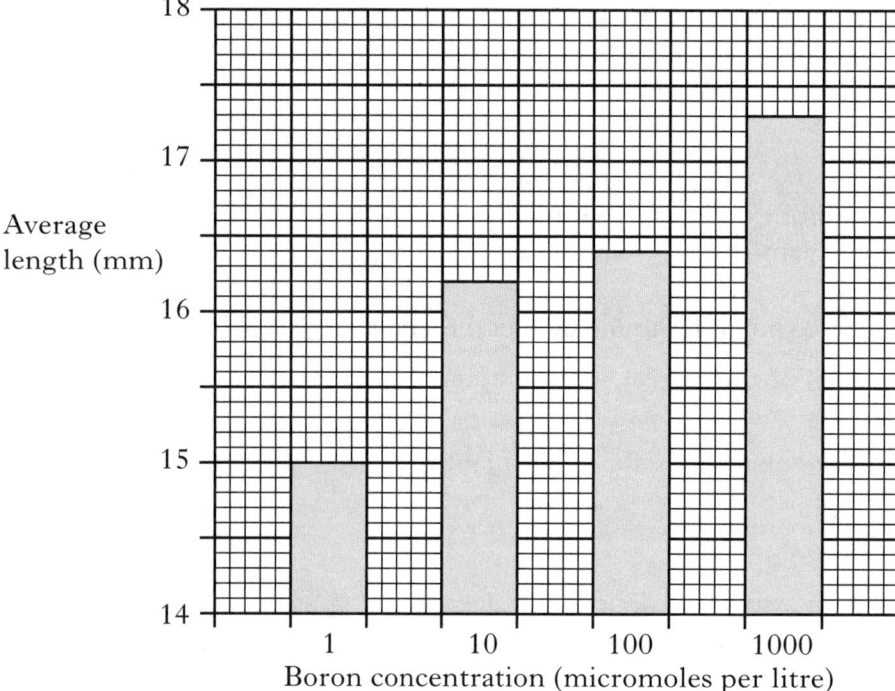

(a) Describe the relationship between boron concentration and the length of the young trout.

_____

_____   **1**

(b) Calculate the percentage change in the average fish length when the boron concentration is increased from 1 micromole per litre to 10 micromoles per litre.

Space for calculation.

_____ %   **1**

DO NOT WRITE IN THIS MARGIN

Marks	KU	PS

7. (continued)

(*c*) Distilled water is the purest form of water available. Give a reason for using distilled water in this investigation.

____

____ **1**

(*d*) Explain why the results would not be valid if the fish were measured more than four weeks after hatching.

____

____ **1**

[Turn over

8. An investigation was carried out into the effect of water concentration on the rate of osmosis.

Details of the apparatus, method used and results are given below.

Apparatus

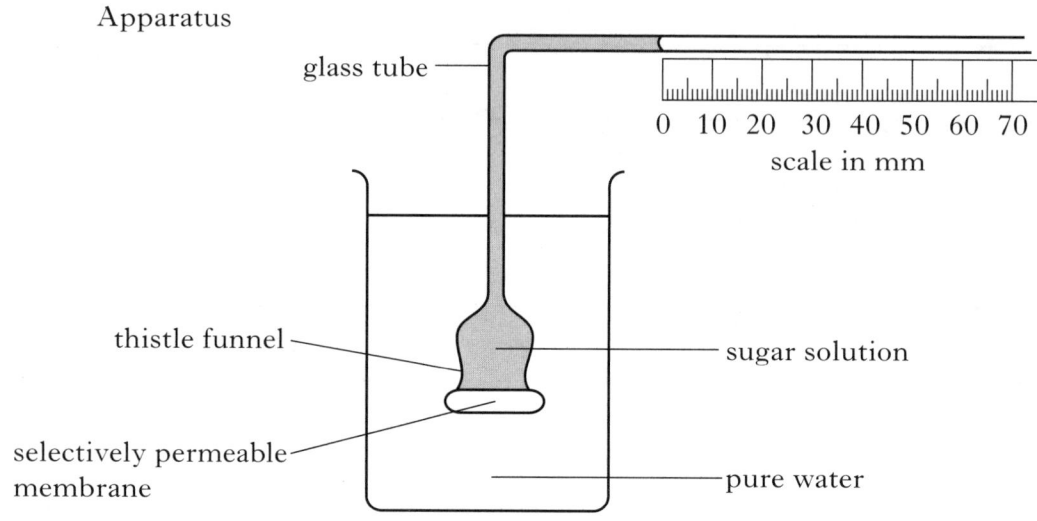

Method

1   A thistle funnel containing 50 cm³ of 0·5% sugar solution was covered with selectively permeable membrane.
2   The funnel was placed in a beaker of pure water.
3   The scale was positioned with the sugar solution at zero on the scale.
4   The position of the sugar solution was recorded after 30 minutes.
5   The procedure was repeated using 1·0%, 2·0% and 3·0% sugar solutions.

Results

Concentration of sugar solution (%)	Distance moved by sugar solution in 30 minutes (mm)
0·5	4·5
1·0	9·0
2·0	18·0
3·0	27·0

Marks | KU | PS

8.   (continued)

(a) Identify **two** variables not already mentioned that should be kept constant when setting up the investigation.

1 _____

2 _____   **2**

(b) Explain the movement of the sugar solution in terms of water concentrations.

_____

_____   **1**

(c) From the results, predict the distance moved by a 3·5% sugar solution in 30 minutes and justify your prediction.

Prediction _____ mm   **1**

Justification _____

_____   **1**

[Turn over

9. (a) The diagram below contains some of the stages of cell division by mitosis.

Describe **Stages 2** and **5** in the spaces provided.

Stage 1

Chromosomes become visible as pairs of identical chromatids.

↓

Stage 2

↓

Stage 3

The spindle fibres contract pulling the chromatids of each chromosome to opposite poles of the cell.

↓

Stage 4

A nuclear membrane forms around each nucleus.

↓

Stage 5

1

1

(b) Mitosis ensures that all daughter cells in a multicellular organism have the same number and type of chromosomes.

Explain why this is necessary.

_____

_____

1

Marks | KU | PS

10. (*a*) Barley is a plant grown for use in the brewing industry. The photographs below show two varieties of barley that have been produced by selective breeding.

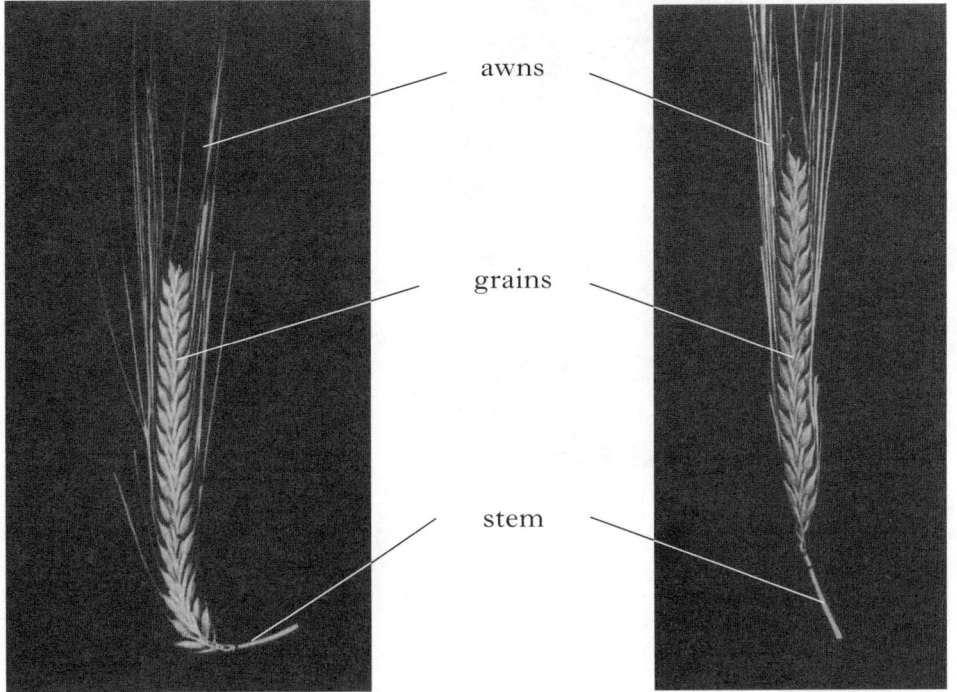

Proctor barley   Rika barley

Describe **one** difference between these two varieties of barley.

_____

_____   **1**

(*b*) (i) Explain why barley must be malted before use in the brewing process.

_____

_____   **1**

(ii) Describe how brewers ensure that the yeast carries out fermentation on the sugars extracted from the malted barley.

_____   **1**

[Turn over

11. (*a*) The photograph shows a child with dimples. Dimples are small indentations in the cheeks. Their presence is controlled by a single gene which has two forms. The dominant form (**D**) gives dimples. The recessive form (**d**) gives no dimples.

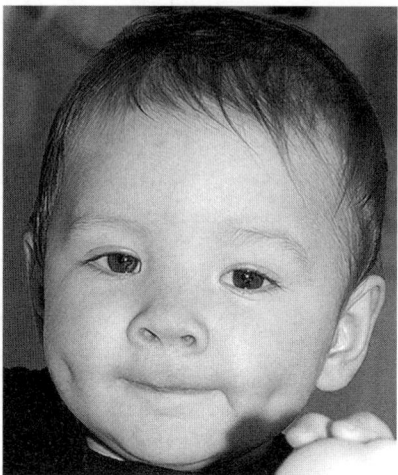

(i) What name is given to different forms of the same gene?

_____   **1**

(ii) The parents of the child are known to have the following genotypes.

DD × dd

Underline **one** option in each bracket to make the following sentence correct.

The parents have $\left\{\begin{array}{l}\text{the same} \\ \text{different}\end{array}\right\}$ phenotypes and

$\left\{\begin{array}{l}\text{the same} \\ \text{different}\end{array}\right\}$ genotypes.   **1**

(iii) What is the genotype of this child?

_____   **1**

DO NOT WRITE IN THIS MARGIN

Marks | KU | PS

11. (continued)

(*b*) The diagram shows a cross between tall and dwarf pea plants.

P **Tall** × **Dwarf**

↓

$\mathbf{F_1}$ all **Tall**

↓

$\mathbf{F_2}$ some **Tall**, some **Dwarf**

(i) What would be the predicted ratio of **Tall** to **Dwarf** plants in the $\mathbf{F_2}$ generation?

______ : ______
Tall **Dwarf** 1

(ii) The observed ratio of **Tall : Dwarf** plants was different from the expected ratio.

Give an explanation for this difference.

____ 1

(iii) Identify the true-breeding plants from the above cross.

Tick (✓) the box(es) of the correct plant(s).

Tall P ☐

Dwarf P ☐

Tall $\mathbf{F_1}$ ☐ 1

[Turn over

Marks | KU | PS

12. An investigation was carried out into the effect of temperature on the rate of respiration by yeast.

Details of the apparatus, method used and results are given below.

Apparatus

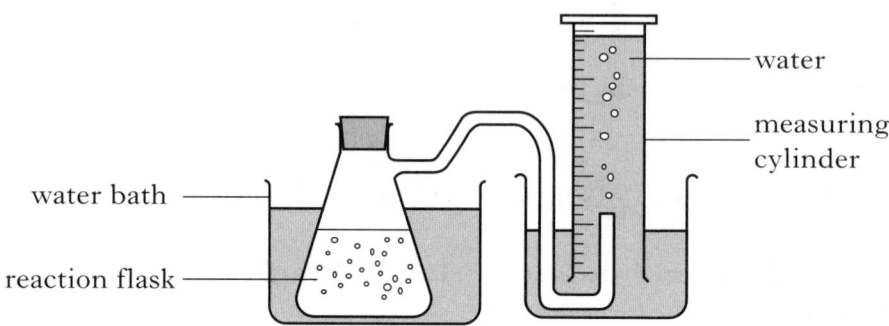

water
measuring
cylinder

water bath

reaction flask

Method

1 Water baths were set up over a range of temperatures.
2 $100\,cm^3$ of glucose solution and $50\,cm^3$ of yeast suspension were allowed to reach the same temperature as the water bath.
3 The glucose solution and the yeast suspension were mixed in the reaction flask.
4 After 1 hour, the volume of gas in the measuring cylinder was measured.

Results

Temperature (°C)	10	20	30	40	50
Volume of gas produced in 1 hour (cm³)	9	18	36	48	5

(a) Ethanol was formed in the reaction flask.

What cell process produced this?

_____   1

(b) Describe the relationship between the temperature and the volume of gas produced in one hour.

_____

_____   2

Marks | KU | PS

12. **(continued)**

(c) Predict the volume of gas which would be collected in one hour if the investigation was repeated at 60 °C. Give an explanation for your answer.

Prediction _____ cm^3

1

Explanation _____

_____

1

(d) Describe the control flasks that would be set up to show that the gas was produced due to activity of the yeast and to no other factor.

_____

_____

2

(e) Use the results to complete a line graph to show the volumes of gas produced in one hour over the range of temperatures.

(An additional grid, if needed, will be found on page 27.)

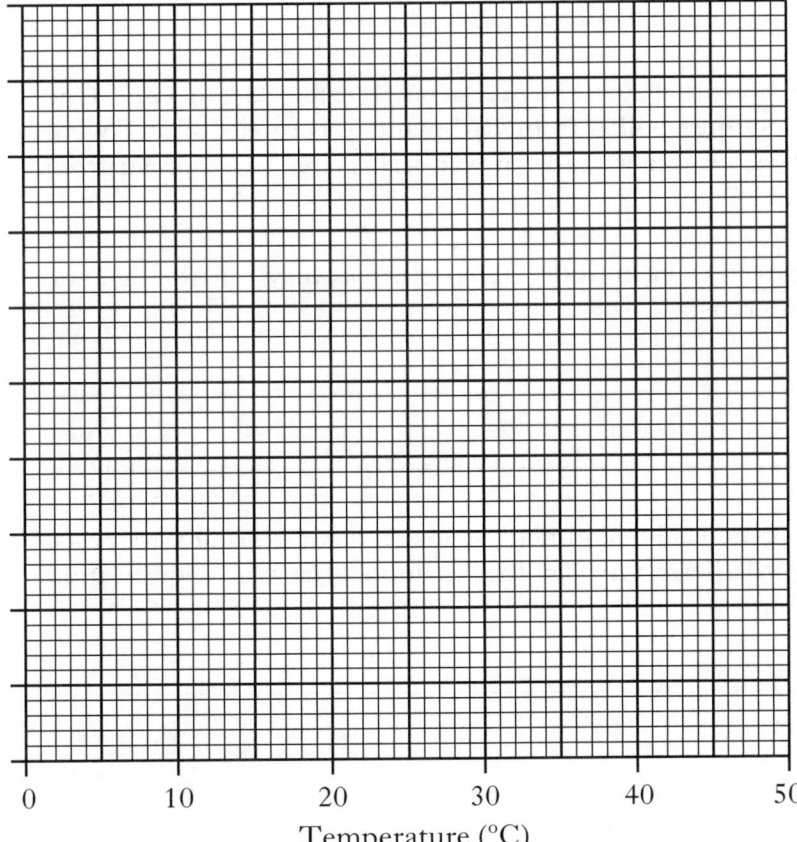

Temperature (°C)

2

[Turn over

DO NOT
WRITE IN
THIS
MARGIN

Marks | KU | PS

13. Read the following passage and answer the questions based on it.

Adapted from **GM Organisms** by John Pickrell, www.newscientist.com

Genetic modification (GM) of crops began with the discovery that the soil bacterium *Agrobacterium* could be used to transfer useful genes from unrelated species into plants. The Bt gene is one of the most commonly inserted. It produces a pesticide toxin that is harmless to humans but is capable of killing insect pests. Many new crop types have been produced. Most of these are modified to be pest, disease or weedkiller resistant, and include wheat, maize, oilseed rape, potatoes, peanuts, tomatoes, peas, sweet peppers, lettuce and onions.

Supporters argue that drought resistant or salt resistant varieties can flourish in poor conditions. Insect-repelling crops protect the environment by minimising pesticide use. Golden rice with extra vitamin A or protein-enhanced potatoes can improve nutrition.

Critics fear that GM foods could have unforeseen effects. Toxic proteins might be produced or antibiotic-resistance genes may be transferred to human gut bacteria. Modified crops could become weedkiller resistant "superweeds". Modified crops could also accidentally breed with wild plants or other crops. This could be serious if, for example, the crops which had been modified to produce medicines bred with food crops.

Investigations have shown that accidental gene transfer does occur. One study showed that modified pollen from GM plants was carried by the wind for tens of kilometres. Another study proved that genes have spread from the USA to Mexico.

(*a*) What role does the bacterium *Agrobacterium* play in the genetic modification of crops?

_____

_____   1

(*b*) Crops can be genetically modified to make them resistant to pests, diseases and weedkillers. Give another example of genetic modification that has been applied to potatoes.

_____   1

DO NOT WRITE IN THIS MARGIN

Marks	KU	PS

13. (continued)

(*c*) Explain why a plant, which is modified to be weedkiller resistant could be:

(i) useful to farmers.

____ **1**

(ii) a problem for farmers.

____ **1**

(*d*) Give **one** example of a potential threat to health by the use of GM crops.

____ **1**

[Turn over

Marks | KU | PS

14. (*a*) In a commercial process, a bacterial species is provided with glucose and produces a hormone. The bacteria release the hormone into surrounding liquid. The graph shows changes in the glucose concentration and the hormone concentration during a 60 hour period.

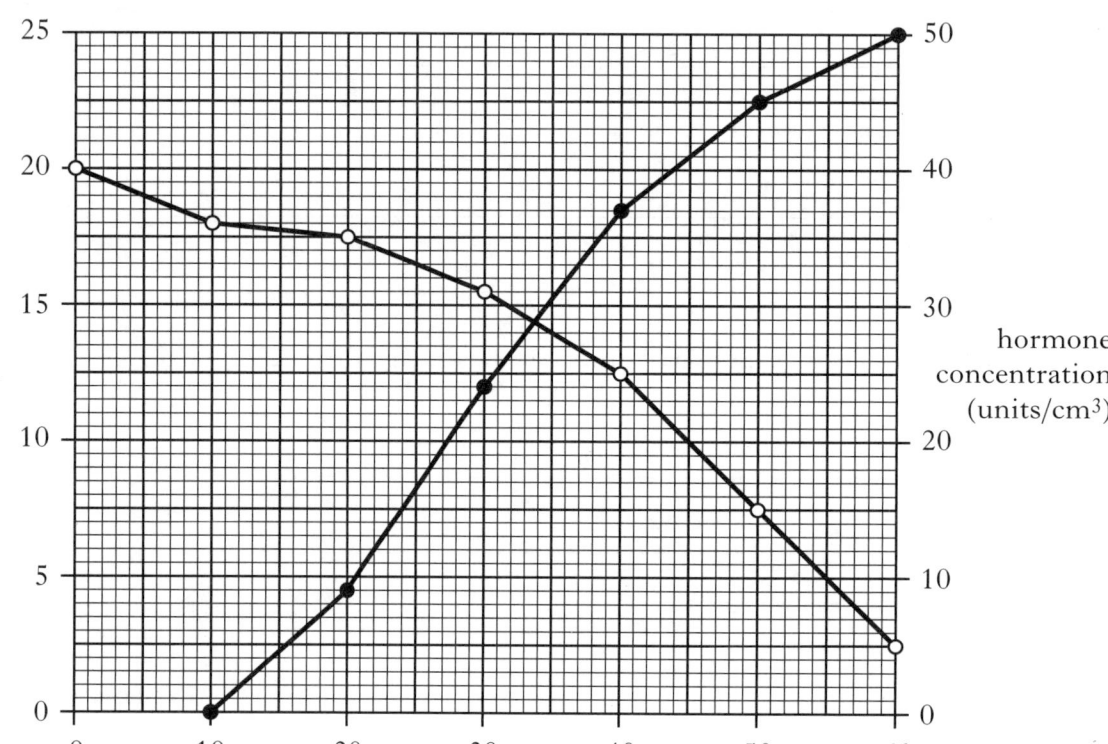

Key ○——○ glucose concentration
         ●——● hormone concentration

(i) How many hours did it take for 50% of the glucose to be used up by the bacteria?

_____ hours

1

(ii) During which 10 hour period was secretion of hormone the greatest?

Tick (✓) the correct box.

☐ 20–30 hours

☐ 30–40 hours

☐ 40–50 hours

☐ 50–60 hours

1

Marks | KU | PS

14. (*a*) (**continued**)

(iii) Calculate the decrease in glucose concentration over the 60 hour period.

Space for calculation.

_____ g/100 cm^3 1

(iv) If glucose continues to be used at the same rate as between 50 and 60 hours, predict how many more hours it would be before all the glucose would be used up.

Space for calculation.

_____ hours   1

(v) During the first 10 hours of the process, energy was being used for functions other than the synthesis of the hormone.

Give **two** pieces of evidence from the graph to support this statement.

1 _____

2 _____   1

(*b*) Glucose is a carbohydrate component of food. Which food component contains most energy per gram?

_____   1

[Turn over for Question 15 on *Page twenty-six*

Marks | KU | PS

15. (*a*) In a sewage works, micro-organisms cause the decay of the sewage. What is the benefit to the micro-organisms in carrying out this process?

_____

1

(*b*) What type of respiration must be carried out by the micro-organisms to ensure complete breakdown of the sewage?

_____

1

(*c*) Sewage contains a wide range of materials. What ensures that all these materials are broken down?

_____

1

(*d*) The table shows the methods of disposal of the sludge obtained from sewage treatment.

Method of disposal of sludge	*Percentage*
Spread on farmland	50
Landfill	10
Dumped at sea	15
Incinerated	20
Other disposal	5

Use the information from the table to complete the pie chart below.

(An additional chart, if needed, will be found on page 27.)

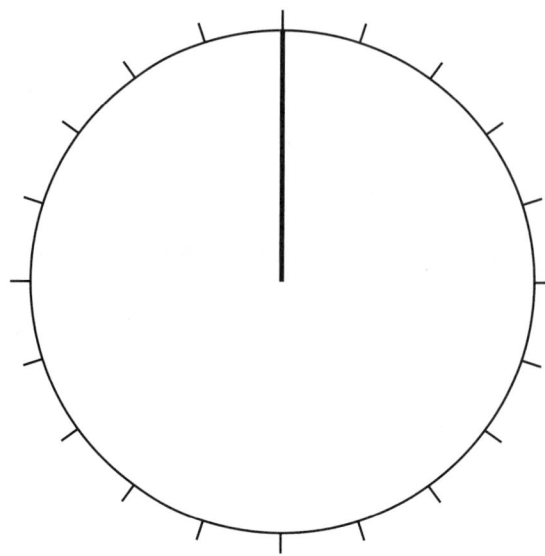

2

ADDITIONAL GRAPH PAPER FOR QUESTION 12(*e*)

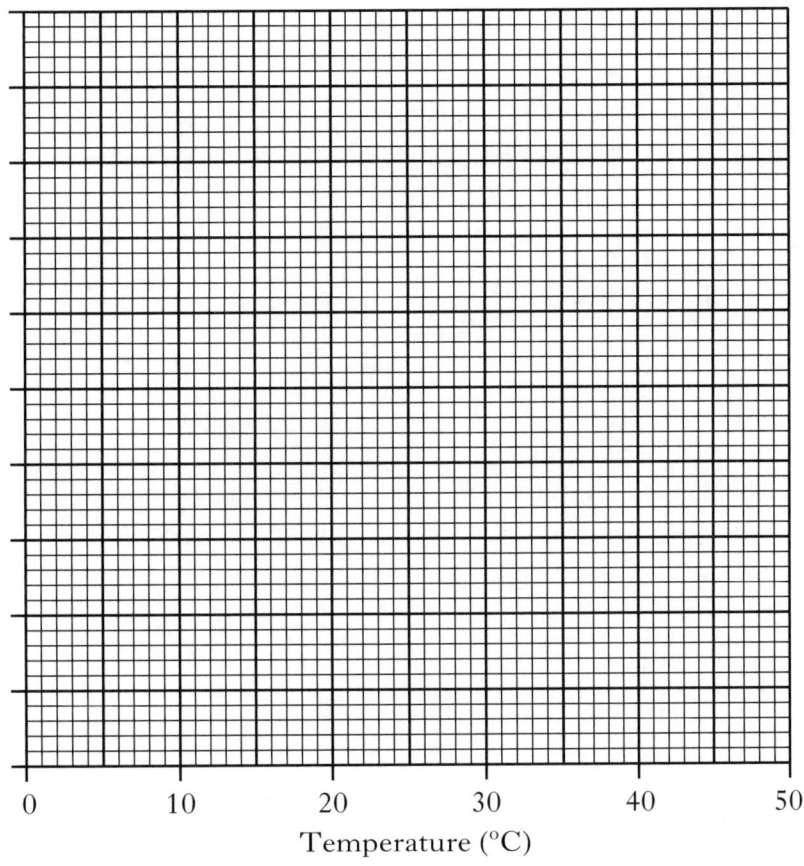

Temperature (°C)

ADDITIONAL PIE CHART FOR QUESTION 15(*d*)

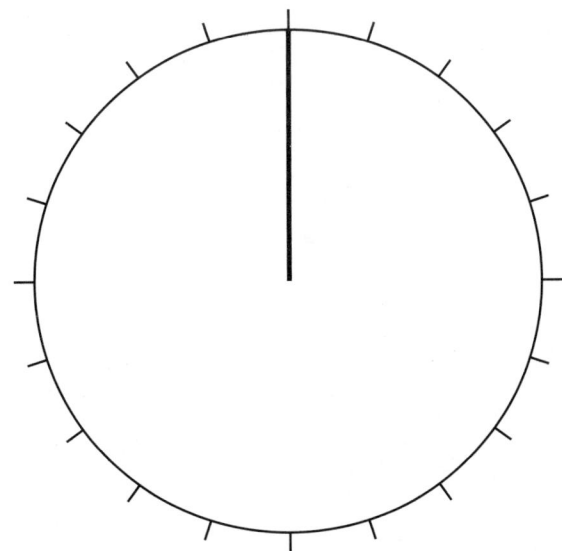

SPACE FOR ANSWERS
AND FOR ROUGH WORKING

STANDARD GRADE | CREDIT

2009

[BLANK PAGE]

FOR OFFICIAL USE

KU   PS

Total Marks

C

0300/402

NATIONAL
QUALIFICATIONS
2009

THURSDAY, 28 MAY
10.50 AM – 12.20 PM

BIOLOGY
STANDARD GRADE
Credit Level

Fill in these boxes and read what is printed below.

Full name o f centre

| |

Town

| |

Forename(s)

| |

Surname

| |

Date of birth
Day Month Year

| | | | | | |

Scottish candidate number

| | | | | | | |

Number of seat

| |

1 All questions should be attempted.

2 The questions may be answered in any order but all answers are to be written in the spaces provided in this answer book, and must be written clearly and legibly in ink.

3 Rough work, if any should be necessary, as well as the fair copy, is to be written in this book. Additional spaces for answers and for rough work will be found at the end of the book. Rough work should be scored through when the fair copy has been written.

4 Before leaving the examination room you must give this book to the invigilator. If you do not, you may lose all the marks for this paper.

Marks | KU | PS

1. (a) Rabbits were first brought to Australia by European settlers.
    The graph below shows the change in rabbit population in Australia since their introduction.

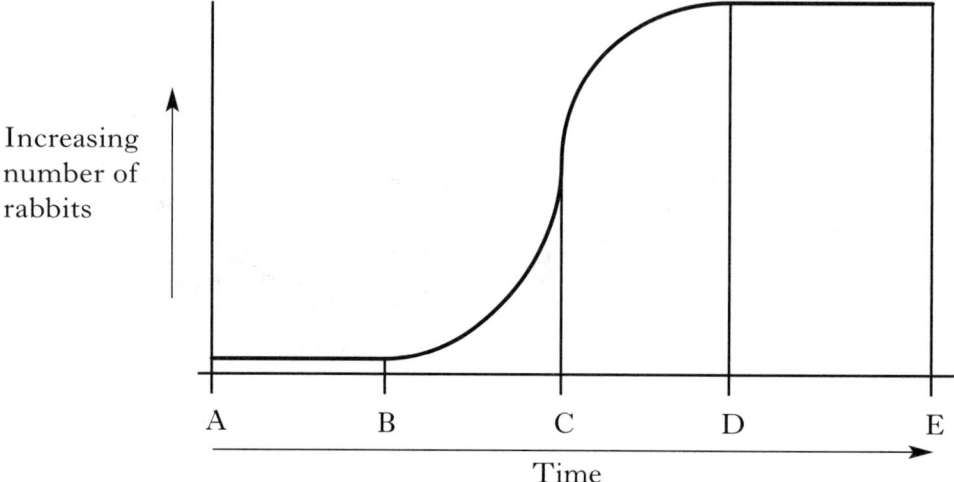

(i) Describe the changes in the rabbit population between times A and E.

_____

_____   **2**

(ii) Suggest one reason for the population change between times B and C.

_____

_____   **1**

(b) To control over-grazing by rabbits, a disease was introduced in 2005 which was fatal to rabbits but not to other species.
    If this disease had wiped out the rabbit population, what effect could it have had on the population of:

(i) Eastern wallabies which are herbivores?

(ii) Dingoes which are carnivorous wild dogs?

Explain your answers.

(i) Effect on Eastern wallabies _____

Explanation _____

(ii) Effect on Dingoes _____

Explanation _____   **2**

Marks | KU | PS

2. (*a*) Coal-burning and nuclear power stations are used to produce electricity in Britain.

Draw lines to connect each type of power station with features considered to be adverse effects of their operation.

Type of power station   *Features*

| coal burning |

- Waste can cause high levels of acid rain

- Waste must be sealed before it is stored

- High volume of greenhouse gas production

| nuclear |

- Waste is dangerous for hundreds of years   **2**

(*b*) Environmental protection analysis was carried out on water samples from three burns.

The Mains Burn had the highest pH at 8·0. It also had the highest oxygen saturation at 94% compared to Bell's Burn which had the lowest at 65%.

The Hatchery Burn had the lowest value for suspended solids at 4·0 mg/l, with an oxygen saturation of 91·5%.

Bell's Burn had a suspended solids reading of 5·6 mg/l and the lowest pH at 7·7 compared to a value of 7·9 for the Hatchery Burn. The highest reading for suspended solids was recorded in the Mains Burn with a value of 6·0 mg/l.

(i) Complete the following table with the data in the passage using suitable column headings.

Analysis site			
Hatchery Burn			
Bell's Burn			
Mains Burn			

3

(ii) Calcium in the water of the burns raises the pH.

Water snails need calcium for shell growth. Which burn would you expect to have the highest number of water snails?

_____   **1**

DO NOT WRITE IN THIS MARGIN

Marks | KU | PS

3. (a) The diagram below represents a wind-pollinated flower.

anther

stigma

Explain how each of the labelled structures contributes to wind pollination.

Anther _____

_____

Stigma _____

_____

2

(b) The chart below shows the peak times for airborne pollen from six wind-pollinated plants.

Type of plant	Month											
---	Jan	Feb	Mar	Apr	May	Jun	Jul	Aug	Sep	Oct	Nov	Dec
Hazel		▨	▨	▨								
Yew		▨	▨	▨								
Willow			▨	▨	▨							
Oil seed rape			▨	▨	▨							
Grass					▨	▨	▨	▨	▨			
Nettle					▨	▨	▨	▨				

(i) How many months are shown to be free of pollen?

_____

1

(ii) The above plants account for most pollen allergy in Britain.

Most allergy sufferers are affected for 3–4 months each year.

Give a conclusion which can be drawn about pollen allergy from these facts.

_____

1

Marks | KU | PS

3. **(b)** **(continued)**

(iii)   In summer, air carries an average of 100 pollen grains per litre.

If a person inhales 12·6 litres of air per minute, calculate the total number of pollen grains inhaled each hour.

Space for calculation

_____ grains per hour   1

(c)   What essential stage in plant reproduction must take place after pollination and before fertilisation?

_____

_____   1

(d)   Give one example of a plant which relies on wind for seed dispersal and describe how its seeds are adapted to dispersal in this way.

Plant _____

Description _____

_____   2

(e)   The list below describes groups of organisms.

1   a patch of strawberry plants produced from the runners of one plant
2   a field of barley grown from seeds
3   a litter of pedigree West Highland Terrier puppies
4   a group of potato tubers harvested from the same plant
5   all the pea plants grown from peas from the same pod

Use the numbers from the list to identify each of the groups which form a clone.

Numbers _____   1

[Turn over

4. (*a*) The table below shows some features of five British butterflies.

Butterfly species	Wing shading	Wing tip	Wing spots
Large White	pale	black	yes
Orange Tip	pale	orange	no
Peacock	dark	blue	yes
Red Admiral	dark	white	yes
Wood White	pale	black	no

Complete the key using the information given in the table.

1 Pale wing shading . . . . go to 2

   Dark wing shading . . . . [_____]

2 [_____] . . . . [_____]

   Orange wing tip . . . . **Orange Tip**

3. Spots on wings . . . . **Large White**

   No spots on wings . . . . [_____]

4. Blue wing tip . . . . **Peacock**

   [_____] . . . . [_____]

3

Marks | KU | PS

4. (continued)

(*b*) The earliest sighting of these butterflies in Britain was recorded in 1956 and again in 2006. The information is shown in the table below.

Butterfly species	Earliest sighting	
	1956	2006
Large White	mid June	early June
Orange tip	late May	mid May
Peacock	mid March	early March
Red Admiral	early June	late May
Wood White	mid May	early May

(i) What evidence suggests that the average temperatures in 2006 were higher than in 1956?

_____

_____   1

(ii) What name is given to organisms, such as these butterflies, which can be used to provide information about environmental factors?

_____   1

[Turn over

Marks | KU | PS

5. (a) The table below shows information on the number of eggs fertilised and the survival of offspring for four different animals.

Animal	Average number of eggs fertilised at one time	Average number of surviving offspring	Percentage survival rate
Dog	5	4	
Human	1	1	100
Bird	4	3	75
Trout	1000	20	2

(i) Calculate the percentage survival rate for the dog and complete the table with the result.

Space for calculation

1

(ii) Explain the difference in the survival rates between humans and trout.

_____

_____

1

(b) Embryos of mammals exchange substances with their mother through the placenta.

Name a substance which passes through the placenta from an embryo to its mother.

_____

1

Marks | KU | PS

6. The diagram below shows *Paramecium*, a single-celled organism which lives in water.

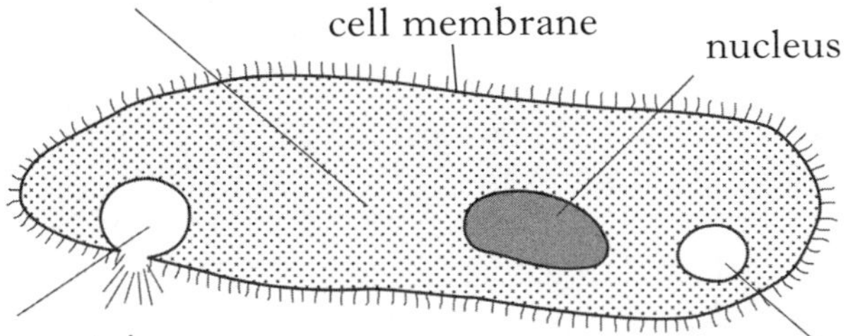

(a) The water concentration outside the cell is higher than the water concentration of the cytoplasm. This causes water to enter the cell constantly.

(i) What is the name for this movement of water?

_____

1

(ii) From the information given, state whether *Paramecium* is likely to live in fresh water or salt water.

_____

1

(b) *Paramecium* must get rid of excess water. Pure water is collected in the vacuoles by removing it from the cytoplasm. The vacuoles are emptied to the surrounding water as soon as they are full.

(i) What would happen to the *Paramecium* cell if the vacuoles stopped working properly?

_____

1

(ii) The vacuoles are not filled by the diffusion of water.

What evidence is there to support this statement?

_____

_____

1

[Turn over

Marks | KU | PS

7. (a) Underline one word in each bracket to make the paragraph about water balance correct.

When a large volume of water is taken into the body, the water

concentration of the blood $\left\{ \begin{array}{c} \text{increases} \\ \text{decreases} \end{array} \right\}$. The volume of ADH

released into the blood by the pituitary gland $\left\{ \begin{array}{c} \text{increases} \\ \text{decreases} \end{array} \right\}$.

This causes water reabsorption by the kidneys to $\left\{ \begin{array}{c} \text{increase} \\ \text{decrease} \end{array} \right\}$

and the volume of urine produced increases.

2

(b) The diagram below represents a nephron from a kidney.

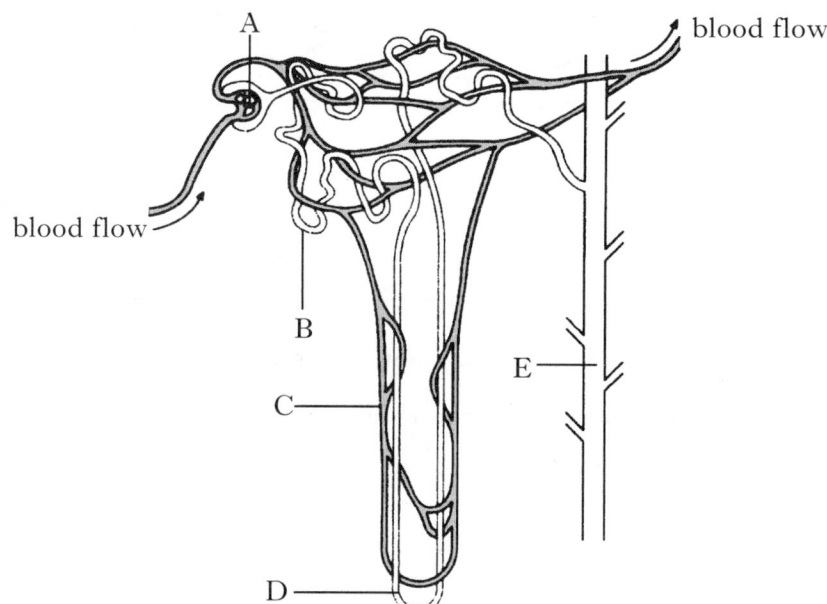

(i) Which letter on the diagram shows where filtration occurs?

_____

1

(ii) Which letter on the diagram shows a collecting duct?

_____

1

7. (continued)

(c) The table below shows the concentration of some substances found in samples taken from the blood, the kidney filtrate and the urine of a volunteer.

Substance	Concentration in blood (g/100cm³)	Concentration in filtrate (g/100cm³)	Concentration in urine (g/100cm³)
urea	0·25	0·25	2·00
glucose	0·10	0·10	0·00
protein	7·50	0·00	0·00
salts	0·62	0·62	1·50

(i) Which substance was present in the blood but was not filtered out of it?

_____

1

(ii) Which substance was filtered from the blood and then completely reabsorbed back into it?

_____

1

(d) A person produces an average of 1·8 litres of urine per day and this is 1% of the kidney filtrate.

What is the average volume of filtrate reabsorbed daily?

Space for calculation

_____ litres

1

[Turn over

DO NOT WRITE IN THIS MARGIN

Marks KU PS

8. (*a*) Stages of mitosis are shown in their correct order in the diagrams below.

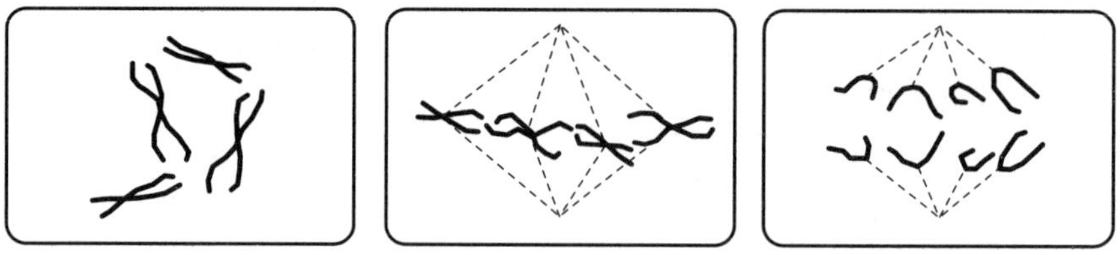

Stage A Stage B Stage C

(i) Label the spindle on one of the diagrams. **1**

(ii) Stage C would be followed by stage D. Describe what would happen in stage D.

____

____ **1**

(*b*) Typical timings of the stages of mitosis are shown in the table below.

Stage	A	B	C	D
Time (minutes)	88	33	25	54

What percentage of the total time for mitosis is taken by stage C?

Space for calculation

________________ % **1**

(*c*) Scientists can grow liver tissue in the laboratory. This is done by making a few liver cells divide by mitosis to form a large mass of cells.

Why is it important that the daughter cells contain the same number of chromosomes as the original mother cells?

____

____ **1**

DO NOT WRITE IN THIS MARGIN

Marks KU PS

9. (*a*) The diagram below represents a hinge joint.

Complete each of the boxes with the missing name or function of the part labelled.

Name	
Function	

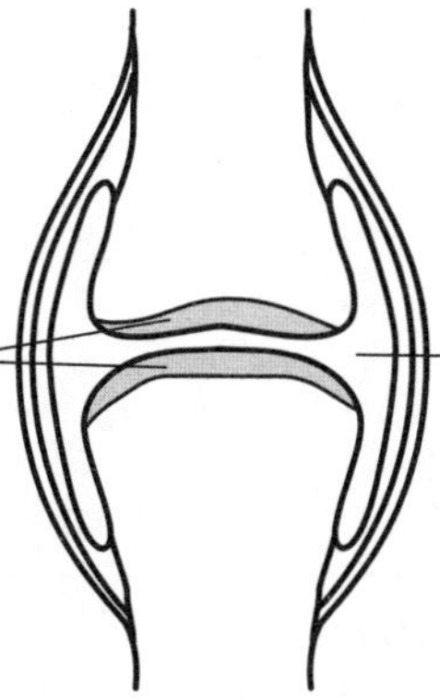

Name	
Synovial fluid	
Function	

2

(*b*) Tendons attach muscle to bone.

Explain why it is important that tendons are inelastic.

____

____

____

2

[Turn over

Marks | KU | PS

10. (*a*) The following statements refer to breathing.

1   ribs move up and out
2   ribs move down and in
3   diaphragm relaxes
4   diaphragm contracts
5   chest volume decreases
6   chest volume increases
7   lung pressure decreases
8   lung pressure increases

Complete the box by inserting the statement numbers which refer to breathing in.

Statements referring to **breathing in**

2

(*b*) The table below shows how exercise at different work rates affects heart rate, breathing rate and the lactic acid concentration in the blood.

Work rate (watts)	Heart rate (beats/min)	Breathing rate (breaths/min)	Lactic acid concentration (mg/l)
0	76	12	1·0
40	92	13	1·5
80	112	15	1·8
120	132	16	3·5
160	156	18	4·5
200	172	30	9·0

(i) Calculate the percentage increase in lactic acid concentration as the work rate increases from 0 to 200 watts.

Space for calculation

_____ %

1

Marks | KU | PS

10. (b) (continued)

(ii) Explain why the lactic acid concentration increases as the work rate increases.

_____

_____ 1

(iii) The graph uses information from the table to show how the breathing rate varies with work rate.

On the same grid, add a scale and label to the vertical axis on the **left side** and plot a line graph to show how the heart rate varies with work rate.

(An additional graph, if needed, will be found on *Page twenty-six.*)

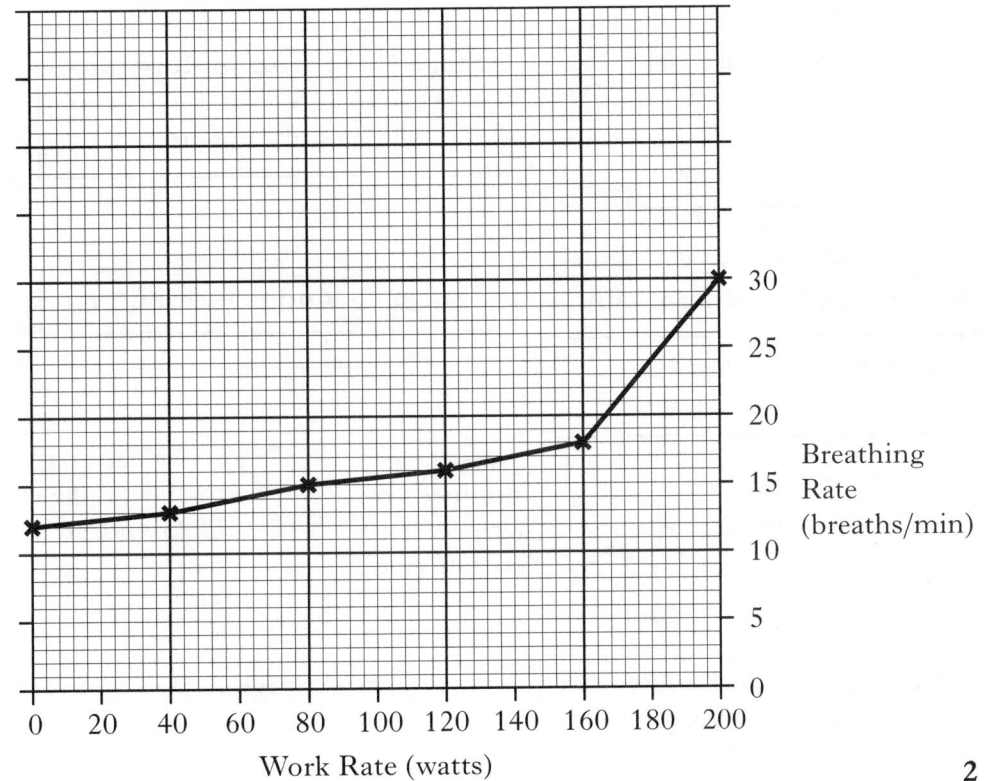

(iv) Describe the relationship between work rate and both breathing and heart rates.

_____

_____ 1

DO NOT WRITE IN THIS MARGIN

Marks | KU | PS

11. The flow chart shows what happens in a typical sewage treatment works.

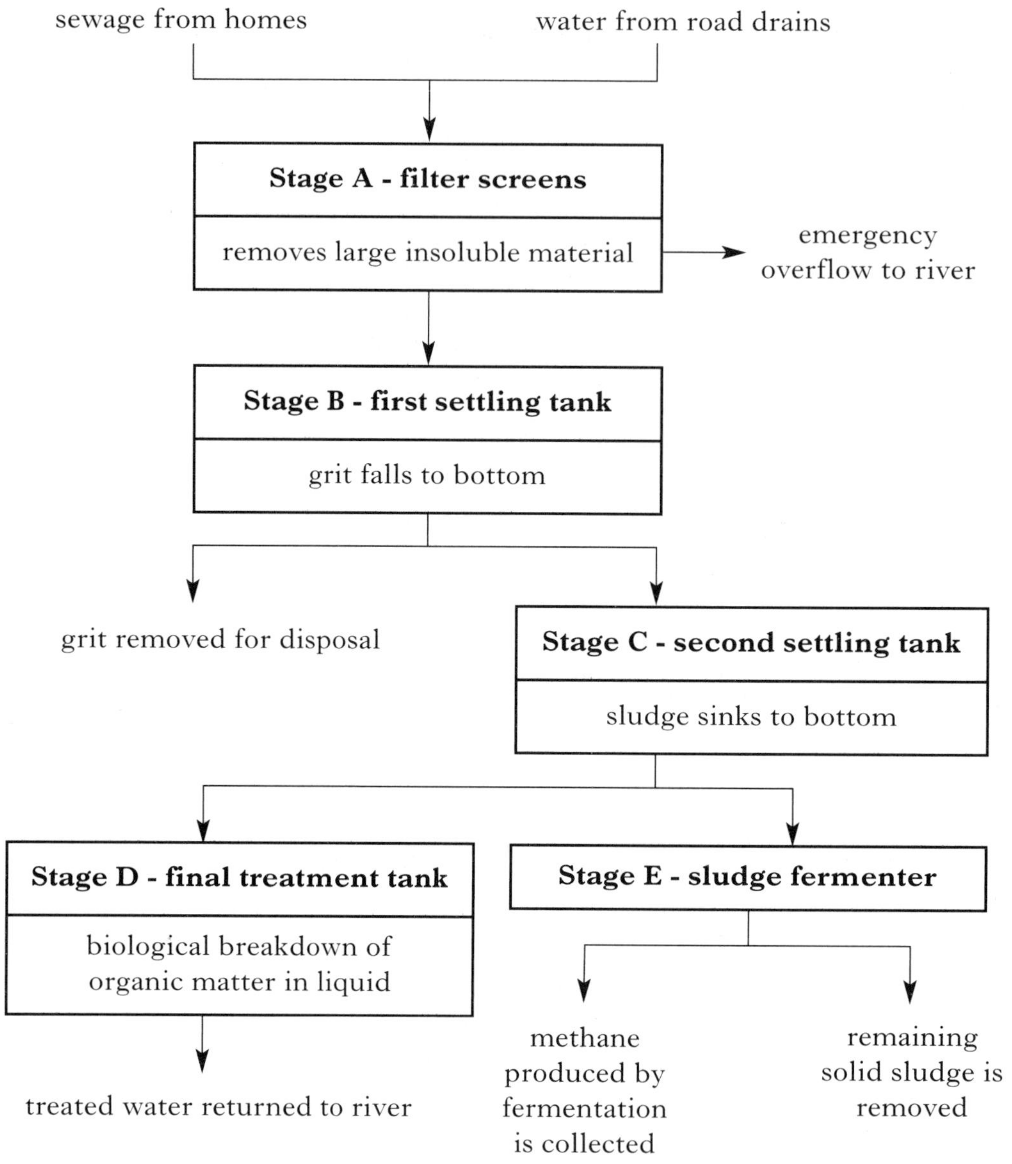
sewage from homes
water from road drains
Stage A - filter screens
removes large insoluble material
emergency overflow to river
Stage B - first settling tank
grit falls to bottom
grit removed for disposal
Stage C - second settling tank
sludge sinks to bottom
Stage D - final treatment tank
biological breakdown of organic matter in liquid
treated water returned to river
Stage E - sludge fermenter
methane produced by fermentation is collected
remaining solid sludge is removed

(*a*) What material, which passes through the screens in **Stage A**, does not reach the tank in **Stage C**?

______________________________ 1

(*b*) Name the gas needed for the final treatment in **Stage D** and explain why the gas is needed for this process.

Gas ______________________________

Explanation ____

____ 2

DO NOT WRITE IN THIS MARGIN

Marks KU PS

11. (continued)

(*c*) When liquid from **Stage D** was sampled, it was found to contain over 80 different species of micro-organisms. Explain why this was seen as a good result.

____

____ **1**

(*d*) Under what environmental conditions could untreated sewage enter the river, even if the sewage treatment plant was working correctly?

____ **1**

[Turn over

12. Read the following passage and answer the questions using information from it.

Salve Imperator   Adapted from "The Life of Birds" by David Attenborough.

Reproduction for Emperor penguins involves extreme hardship. They start their breeding cycle in March at the beginning of the Antarctic winter. At this time the fringe of ice that surrounds the Antarctic continent is at its narrowest. The penguins walk across it for several miles to the permanent ice which is their breeding ground. Up to 25 000 penguins gather and mating takes place in April.

As the temperature falls, the sea ice expands by 2 miles per day. In May the female produces one large egg which she places on the top of her feet. The male takes the egg, juggles it onto the top of his feet and covers it with a fold of his densely feathered abdomen to keep it warm. Producing the egg has taken a significant proportion of the female's body reserves. She needs to replenish them urgently and heads back to sea to feed.

As the winter winds begin to blow, the temperature falls. The male Emperors huddle closer together for warmth and shelter. They use their tiny stump of a tail as the third leg of a tripod and rest on their heels. Their upwardly turned toes keep their precious eggs off the ice. There is nothing to eat and for a month there is total darkness.

After 60 days the eggs hatch. The males, close to starvation, manage to produce a little milky secretion from their gullets for their chicks. At this critical moment the females reappear. They have had a long journey as the ice has extended considerably. The females regurgitate their chicks' first real meal. The males now start the long trek back to the sea to feed for the first time in four months.

Three weeks later, the males are back to take over the care of the chicks, allowing the females to return to the sea. As winter slackens its grip, the ice begins to break up. The journey to the sea gets shorter and the parents can increase the frequency of feeding. In November the parents stop feeding the young and long processions of adults and young waddle down to the sea.

(a) Why is it necessary for the females to leave their eggs and return to the sea?

_____

1

Marks | KU | PS

12.   (continued)

(*b*)   By how much has the distance to the sea increased in the time between laying and hatching?

Space for calculation

_____ miles

1

(*c*)   How does a male keep his egg off the ice?

_____

_____

1

(*d*)   The following list describes events in the life of Emperor penguins.

List   1   walk to breeding grounds
              2   mating
              3   egg laying
              4   eggs hatch and females return
              5   parents and chicks waddle to the sea

Complete the time line below by placing the number of each event in the correct month.

(An additional time line will be found, if needed, on *Page twenty-six*).

Time line

Jan	Feb	Mar	Apr	May	Jun	Jul	Aug	Sep	Oct	Nov	Dec

2

(*e*)   How many months of the year are **not** spent breeding and rearing young?

_____

1

[Turn over

Marks | KU | PS

13. The following apparatus was used to investigate the effectiveness of washing powders.

Identical pieces of stained cloth were washed using different washing powders.

The cloths were dried and the degree of stain removal was measured by recording light reflected from the cloth with a light meter. The meter was set to read 100% when the cloth was perfectly clean. Any stain left on the cloth reduced the intensity of light recorded.

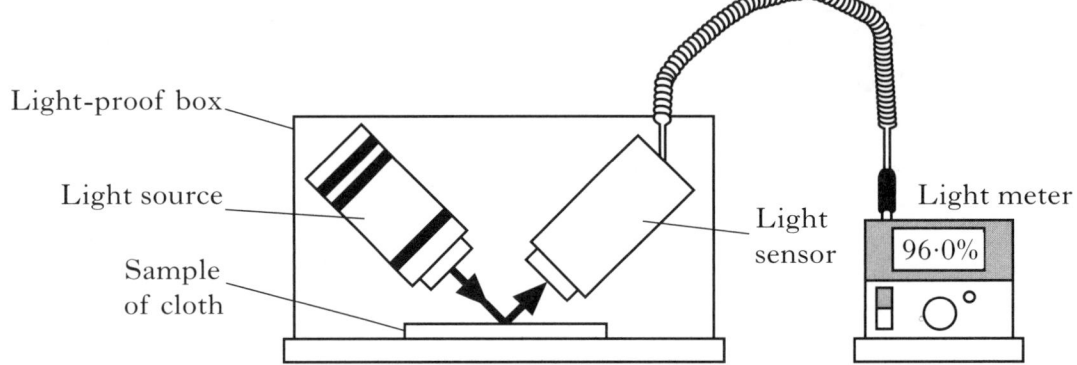

(a) (i) Various precautions were taken to ensure that the experimental procedure was valid.

Identify the point(s) which contributed to this.

Tick (✓) the correct box(es).

The procedure used gave appropriate information about the effectiveness of washing powders. ☐

All significant variables were controlled and were identical except the one being investigated. ☐

Several results were collected and used to calculate an average. ☐

1

(ii) Explain why it was necessary to carry out the investigation in a light-proof box.

_____

_____

1

Marks | KU | PS

13. (continued)

(b) The results obtained using two different washing powders at various temperatures are shown below.

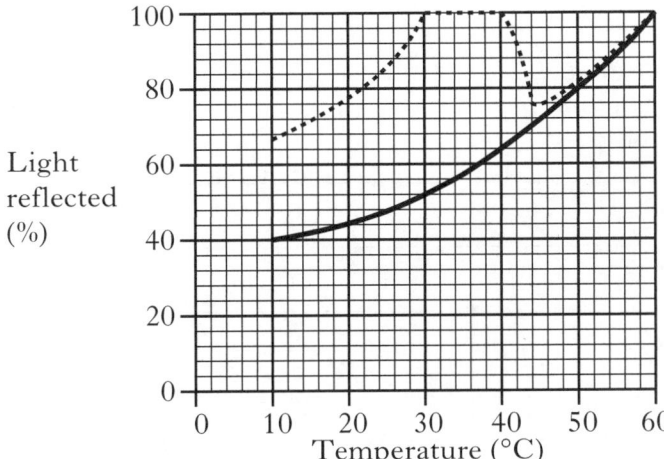

Key

·········· Biological

———— Non-biological

(i) At which temperature was there the greatest difference between the effectiveness of the two washing powders?

_____ °C

1

(ii) Each one degree Celcius reduction in the washing temperature saves 2p in the cost of electricity used to heat the water for each wash.

Calculate the annual saving in the electricity costs to achieve 100% stain removal with biological washing powder compared to a non-biological one, for a household which does one wash per week.

Space for calculation

annual saving = £ _____

1

(iii) What type of biological substance gives biological washing powders their properties?

_____

1

(iv) Explain why the effectiveness of the biological washing powder decreases between 40°C and 45°C.

_____

1

[**Turn over**

Marks | KU | PS

14. Micro-organisms living in water use dissolved oxygen for respiration.

The mass of oxygen they use is called the Biochemical Oxygen Demand (BOD).

The table below shows the BOD of a river and the concentration of solid material carried by the river during five months of the year.

Month	Concentration of solid material (mg/l)	BOD (mg/l)
January	6·75	1·0
March	7·25	1·2
May	10·75	1·9
September	5·50	0·5
November	9·00	1·5

(a) Use the information in the table to complete the bar chart below for January and November.
(An additional chart, if needed, will be found on *Page twenty-seven*.)

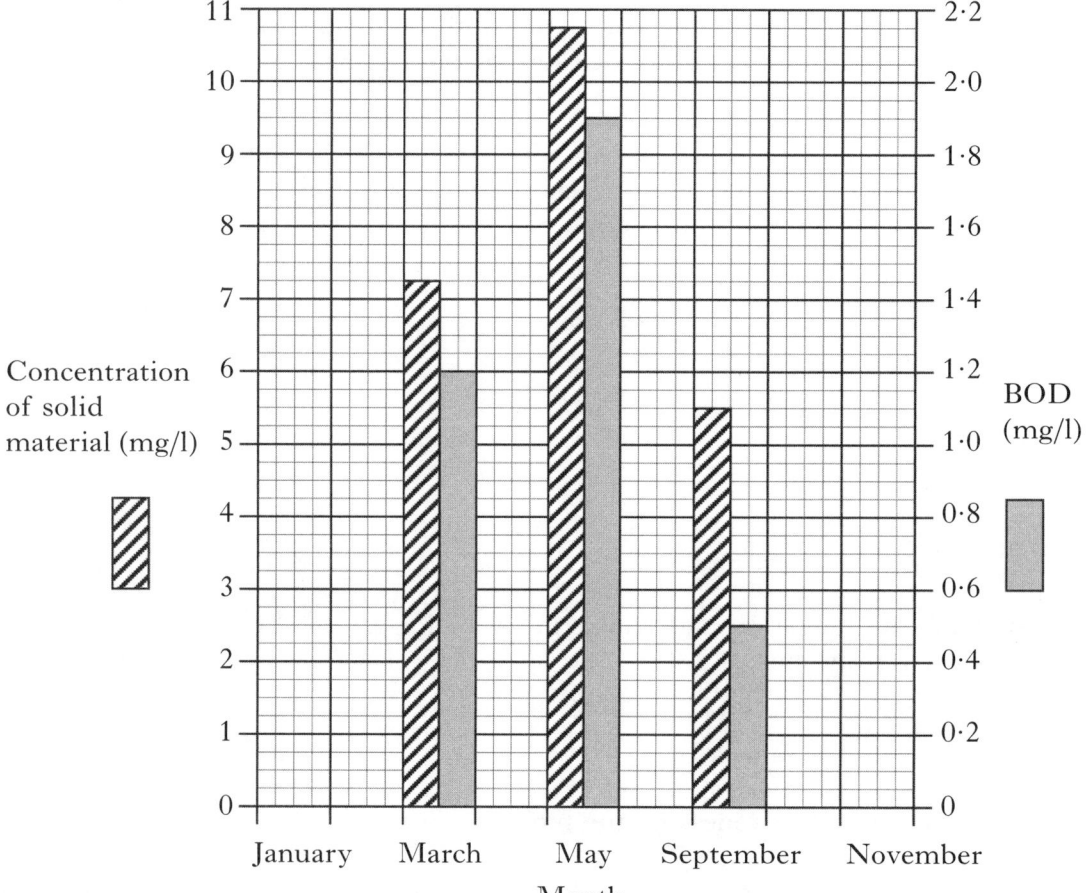

1

DO NOT
WRITE IN
THIS
MARGIN

Marks | KU | PS

14.   **(continued)**

(b) Describe the relationship between the concentration of solid material in the river water and the BOD.

_____

_____

1

(c) After heavy rains in December, the concentration of solid material in the water was found to be 10·0 mg/l.

What would be the expected BOD for this sample?
Tick (✓) the correct box.

7·5 mg/l   ☐

5·0 mg/l   ☐

1·75 mg/l   ☐

1·25 mg/l   ☐

1

[Turn over

DO NOT WRITE IN THIS MARGIN

Marks | KU | PS

15. Candytuft is a plant with white or pink flowers. The two forms of the gene responsible for the flower colour are:

P = pink flowers   and   **p** = white flowers.

(a) A plant breeder crossed two pink flowered plants as shown below.

Parents   **Pp**   ×   **Pp**

(i) What is the expected ratio of pink to white flowered plants in the offspring?

_____ : _____

   pink : white

1

(ii) If 48 offspring had been produced, how many white flowered plants would have been expected?

Space for calculation

_____ white flowered plants

1

(iii) The offspring actually consisted of 24 pink flowered and 16 white flowered plants.

What is the simplest whole number ratio of pink to white flowered plants in the offspring?

Space for calculation

_____ : _____

   pink : white

1

(iv) Suggest a reason for the difference between the expected ratio and the observed ratio.

_____

DO NOT WRITE IN THIS MARGIN

Marks KU PS

15. (continued)

(*b*) What name is given to two different forms of a gene?

______________________________ **1**

(*c*) Some plant characteristics show discontinuous variation. What is meant by "discontinuous variation"?

____

____ **1**

[*END OF QUESTION PAPER*]

ADDITIONAL GRAPH FOR QUESTION 10 (*b*) (iii)

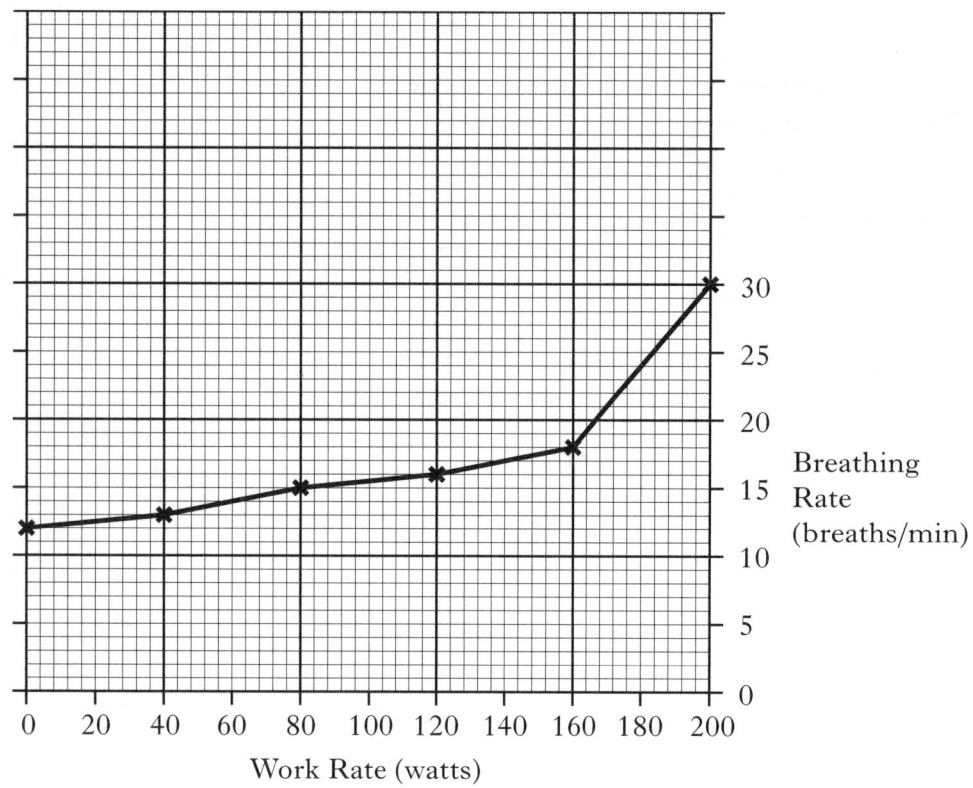

Breathing Rate (breaths/min)

Work Rate (watts)

ADDITIONAL TIME LINE FOR QUESTION 12 (*d*)

Time line

Jan	Feb	Mar	Apr	May	Jun	Jul	Aug	Sep	Oct	Nov	Dec

ADDITIONAL CHART FOR QUESTION 14 (*a*)

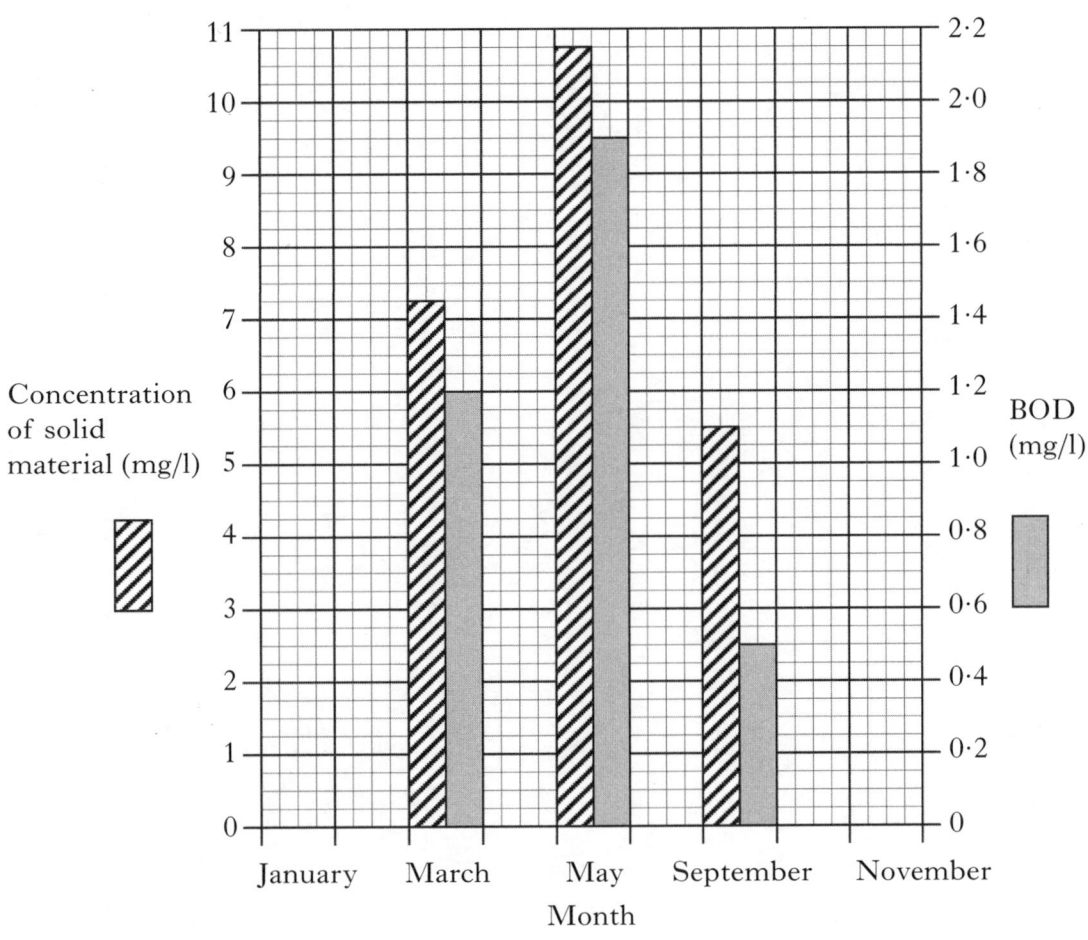

SPACE FOR ANSWERS
AND FOR ROUGH WORKING

STANDARD GRADE | CREDIT

2010

[BLANK PAGE]

C

FOR OFFICIAL USE

KU	PS

Total Marks

0300/402

NATIONAL
QUALIFICATIONS
2010

THURSDAY, 27 MAY
10.50 AM – 12.20 PM

BIOLOGY
STANDARD GRADE
Credit Level

Fill in these boxes and read what is printed below.

Full name of centre

Town

Forename(s)

Surname

Date of birth

Day	Month	Year	Scottish candidate number	Number of seat

1   All questions should be attempted.

2   The questions may be answered in any order but all answers are to be written in the spaces provided in this answer book, and must be written clearly and legibly in ink.

3   Rough work, if any should be necessary, as well as the fair copy, is to be written in this book. Additional spaces for answers and for rough work will be found at the end of the book. Rough work should be scored through when the fair copy has been written.

4   Before leaving the examination room you must give this book to the Invigilator. If you do not, you may lose all the marks for this paper.

Marks   KU PS

1. (a) Two groups of pupils set pitfall traps in the school gardens to sample invertebrates living there. All traps were left for the same length of time. The results are shown in the following tables.

Group A	Pitfall trap number	Number of each type of invertebrate caught				
		spider	beetle	snail	earthworm	woodlouse
	1	2	1	2	0	1
	2	3	2	1	0	0

Group B	Pitfall trap number	Number of each type of invertebrate caught				
		spider	beetle	snail	earthworm	woodlouse
	1	2	3	2	1	1
	2	2	0	3	1	2
	3	0	2	1	1	1
	4	3	2	1	0	1
	5	3	1	1	2	1

(i) How many types of invertebrate did Group A find?

_____ types   1

(ii) Calculate the average number of spiders found in Group B's traps.

Space for calculation

_____ spiders   1

(iii) Explain why conclusions made by Group B from their results would be more reliable than conclusions made by Group A.

_____

_____   1

(iv) Give **one** precaution which must be taken when setting up a pitfall trap, or other named sampling technique, and explain the reason for it.

Sampling technique _____

Precaution _____

_____   1

Reason _____

_____   1

1. **(continued)**

(b) The diagrams below show the invertebrates collected by the pupils.

They are not drawn to scale.

Earthworm

Snail

Spider

Beetle

Woodlouse

(i) Complete the following key using information from the diagrams.

1   Legs ....   Go to 2

     No legs ....   Go to ☐   **1**

2   12 legs or more ....   *Woodlouse*

     Fewer than 12 legs ....   Go to 3

3   Spots on body ....   *Beetle*

     No spots on body ....   ☐   **1**

4   Shell ....   *Snail*

     ☐ .... ☐   **1**

(ii) Give **three** features of the beetle mentioned in the key.

1 _____

2 _____

3 _____   **1**

DO NOT
WRITE IN
THIS
MARGIN

Marks   KU | PS

2. (a) Electricity can be generated by using fossil fuels or nuclear fuels as energy sources.

Give **one** disadvantage of using each type of fuel.

Fossil fuel _____

_____   1

Nuclear fuel _____

_____   1

(b) (i) Micro-organisms can obtain their energy by feeding on organic waste such as sewage.

Explain why each of the following events occurred after raw sewage was accidentally released into a river.

1   The number of micro-organisms in the river increased.

_____

_____   1

2   The number of fish in the river decreased.

_____

_____   1

(ii) A group of students monitored the river using indicator species.

What is meant by the term "indicator species"?

_____

_____   1

DO NOT
WRITE IN
THIS
MARGIN

Marks | KU | PS

3. (*a*) An investigation was carried out into the effect of temperature on the germination of grass seeds.

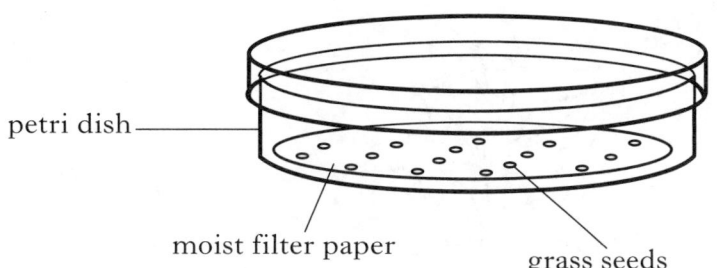

petri dish

moist filter paper

grass seeds

Five identical petri dishes, each containing 20 seeds, were set up as shown in the diagram. Each dish was left in the dark at a different temperature. After seven days the percentage germination in each dish was calculated. The results are shown in the table below.

Temperature (°C)	10	18	27	36	45
Percentage germination	45	65	80	70	40

(i) From the results, what is the optimum temperature for the germination of this species of grass?

_____ °C

1

(ii) Name **one** factor, not already mentioned, which should be kept the same for all the dishes.

_____

1

(iii) What feature of the investigation was designed to increase the reliability of the results?

_____

1

(*b*) Describe the changes in the percentage germination of seeds that occur over a range of temperatures.

_____

_____

2

[Turn over

4.   Rooting compound helps plant cuttings to produce new roots. The diagram below shows the apparatus used to find out how the concentration of rooting compound affects this.

Six flasks were set up, each with a different concentration of rooting compound.

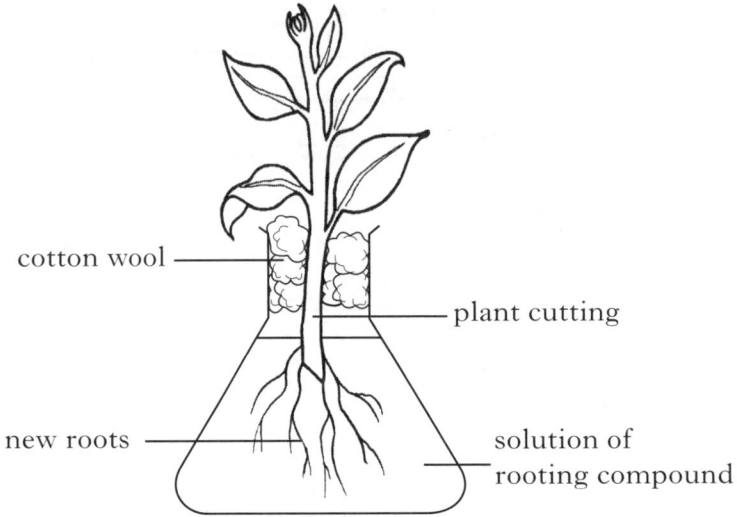

cotton wool

plant cutting

new roots

solution of rooting compound

After 21 days the number of roots and the lengths of the roots on each cutting were measured.

The results are shown on the following graph.

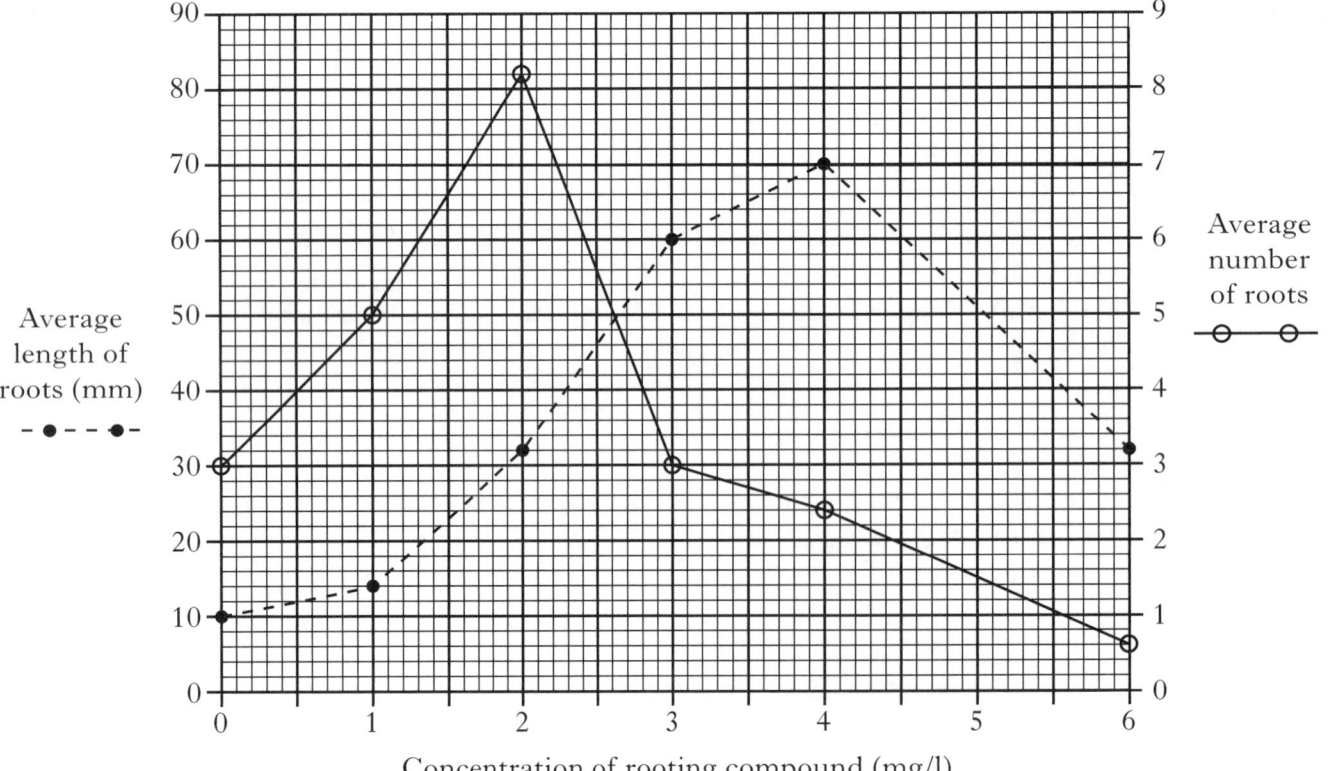

Marks

4. **(continued)**

(a) (i) Which **two** concentrations of rooting compound, used in the investigation, produced the same average root length?

_____ mg/l and _____ mg/l

1

(ii) Using information from the graph, predict the average length of roots on cuttings grown in a concentration of 2·5mg/l.

_____ mm

1

(iii) Which concentration of rooting compound produces the greatest number of roots per cutting?

_____ mg/l

1

(iv) Describe how the average length of the roots on one cutting would be calculated.

_____

_____

_____

1

(b) Give **one** advantage to a gardener of producing plants from cuttings rather than from seeds.

_____

1

(c) What term is given to a group of plants grown from cuttings taken from a single plant?

_____

1

[Turn over

5. (*a*) The following table gives information about reproduction in various animals.

	Average number of eggs or young produced per year	*Type of fertilisation*	*Where development takes place*
cod	6 million	external	water
frog		external	water
blackbird	5	internal	inside eggshell
stoat	4	internal	inside female

(i) A female frog produces a total of 4000 eggs over a five year period.

1 Complete the table to show the average number of eggs she produces per year.

Space for calculation

1

2 On average, two eggs from each female frog must survive to breeding age to keep the population constant. What percentage of this frog's **total** egg production does this represent?

Space for calculation

_____ %

1

(ii) Explain why fish such as cod must produce far more eggs than mammals such as stoats to ensure the survival of the species.

_____

_____

1

(iii) Explain the importance of internal fertilisation to land-living animals.

_____

_____

1

Marks | KU | PS

5. **(continued)**

(*b*) The diagram below represents a stage in the development of a human fetus.

X

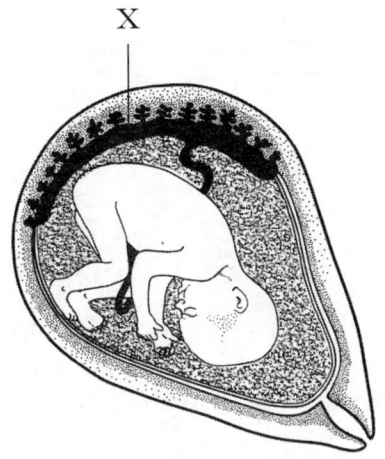

Name structure X and give **one** of its functions.

Name _____

Function _____

_____   **2**

[Turn over

Marks KU PS

6. The apparatus shown below was used to study the effect of different temperatures on the activity of the enzyme catalase.

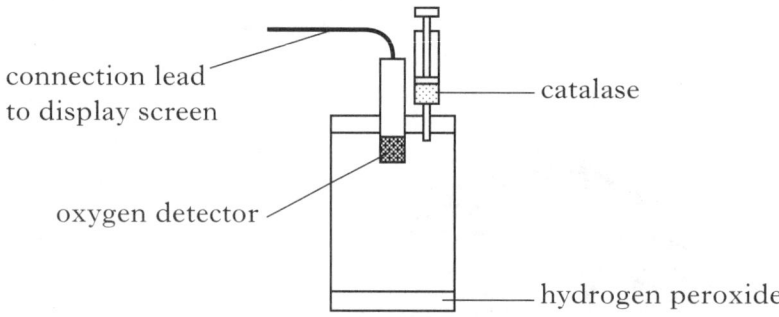

The catalase was added and reacted with the hydrogen peroxide to release oxygen. The increase in oxygen compared to the starting value was recorded as a percentage.

This was carried out at five different temperatures and the results are shown below.

Temperature (°C)	Increase in oxygen (%)
4	0·55
21	0·80
34	1·45
40	1·05
50	0·05

(a) Use the results to draw a line graph.

(An additional grid, if needed, will be found on *Page twenty-three*.)

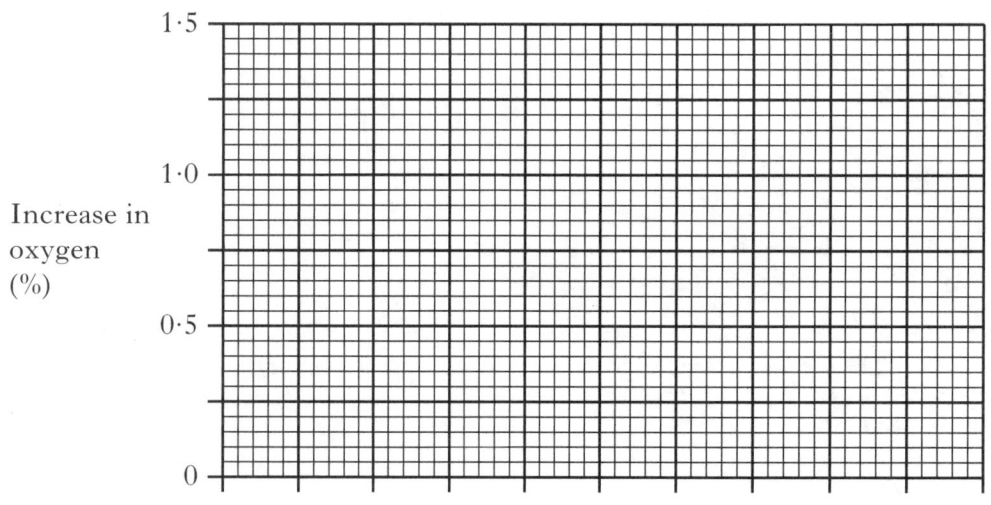

2

DO NOT WRITE IN THIS MARGIN

Marks | KU | PS

6. (continued)

(b) At which temperature was the catalase most active?

_____ °C

1

(c) Why was it important that the catalase and the hydrogen peroxide were both at the required temperature before the catalase was added?

_____

_____

1

(d) Explain why there was no oxygen released when the experiments were repeated with different enzymes.

_____

_____

1

(e) Calculate the simple whole number ratio of percentage increase in oxygen at 34 °C, 40 °C and 50 °C.

Space for calculation

_____ : _____ : _____
34 °C   40 °C   50 °C

1

[Turn over

Marks

7. The diagrams below represent red blood cells in different solutions as they would appear under a microscope.

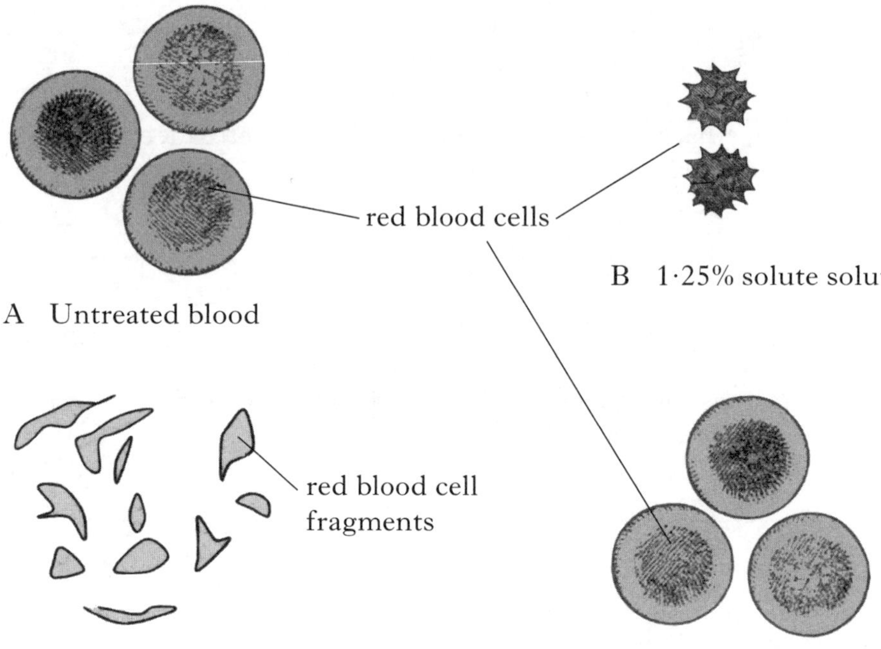

red blood cells

A Untreated blood

B 1·25% solute solution

red blood cell fragments

C 0·25% solute solution

D 0·90% solute solution

(a) Use the information in the diagrams to predict the percentage solute concentration of human blood. Explain your answer.

Solute concentration _____ %

Explanation _____

_____

_____ 1

(b) What has happened to the cells in diagram B? Explain the change in terms of water concentrations.

Description _____

Explanation _____

_____ 2

Marks

8. The diagram below represents part of a finger joint.

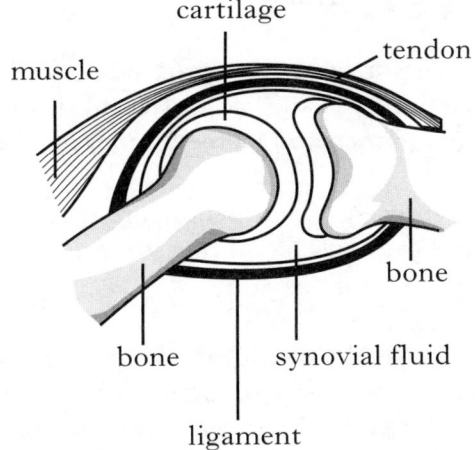

(a) (i) The joint needs a second muscle and tendon to make it function properly. Explain the need for joints to have muscles which work in pairs.

_____

_____

_____   2

(ii) What feature of tendons ensures that all the force from a muscle contraction is transmitted to the bone?

_____   1

(b) Name **two** parts of the joint which reduce friction.

1 _____

2 _____   1

[Turn over

Marks | KU | PS

9. Read the following passage and answer the questions based on it.

Young at Heart?

New research shows that decades of hard-won progress in reducing the risk of heart disease in America appears to be losing pace. Recent death rates from heart disease remain almost unchanged in men and women under 55 years old.

This trend comes at a time when even young people are increasingly likely to be obese, suffer from diabetes and have high blood pressure. Each of these increases heart attack risk.

Data from 1980 to 2002 showed that the death rate from heart disease had fallen. In the whole population there was a yearly reduction of 2·9 percent during the 1980s, 2·6 percent during the 1990s and 4·4 percent from 2000 to 2002.

However the numbers told a strikingly different story for people aged 35 to 54. The yearly death rate from heart disease fell by 6·2 percent in the 1980s, by only 2·3 percent in the 1990s and showed no reduction at all between 2000 and 2002.

The message is that heart disease has not gone away, and could become an even greater problem if people fail to pay attention to known warning signs. Dr F S Ford, a medical officer for the American government said, "Young adults should take stock of their lifestyles. Don't smoke and take at least 30 minutes of exercise per day. If you need to lose weight, you must burn more energy than you take in. Good habits should start early. Changes that lead to heart disease, for example hardening of the arteries, occur at an early age. Therefore it is especially important that children and young people develop appropriate habits that minimise their risk of heart disease later in life."

(*a*) From the passage, identify **three** factors which contribute to the risk of heart disease.

1 _____

2 _____

3 _____   1

(*b*) Complete the table below to show the changes in death rates for the whole population and for the 35–54 age group.

	Average yearly reduction in death rate from heart disease (%)		
	1980–1989	1990–1999	2000–2002
Whole population			
35–54 age group			

2

DO NOT WRITE IN THIS MARGIN

Marks | KU | PS

9. (continued)

(*c*) According to Dr Ford, why is it important that "good habits should start early"?

____

____ **1**

(*d*) What cellular process is being referred to in the phrase "you must burn more energy"?

________________________ **1**

[Turn over

10.   A tin containing 170 g of evaporated milk has the following label.

> ### Typical values per tin
>
> | | |
> |---|---|
> | Energy | 1156 kJ |
> | Protein | 12·75 g |
> | Carbohydrate | 17·47 g |
> | Fat | 17·45 g |
> | Fibre | 0·00 g |
> | Salt | 0·33 g |

(a)   (i)   What percentage of the total contents of the tin is protein?

Space for calculation

_____ %   1

(ii)   What component of the milk would provide most energy?

_____   1

(b)   Name the chemical elements present in fats.

_____   1

11. (a) <u>Underline</u> **one** option in each bracket to make the following sentence about breathing correct.

When breathing out, the lung volume $\left\{ \begin{array}{l} \text{increases} \\ \text{stays the same} \\ \text{decreases} \end{array} \right\}$ and as a result the

air pressure in the lungs $\left\{ \begin{array}{l} \text{increases} \\ \text{stays the same} \\ \text{decreases} \end{array} \right\}$.

1

(b) The effect of changing the carbon dioxide concentration in inhaled air on a person's breathing was investigated.

The table below shows the average volume of air inhaled each minute at different concentrations of carbon dioxide.

Carbon dioxide concentration in inhaled air (%)	0	2	4	6	8
Average volume of air inhaled (litres per minute)	8	12	16	24	60

(i) How many times greater is the average volume of air inhaled per minute when the carbon dioxide concentration is increased from 2% to 8%?

Space for calculation

_____ times

1

(ii) Calculate the average volume of carbon dioxide entering the lungs each minute when the carbon dioxide concentration in the air is 4%.

Space for calculation

_____ litres

1

(iii) Calculate the increases in the average volume of air breathed per minute when the carbon dioxide changes from 0 to 2% and from 6 to 8%.

Express these increases as a simple whole number ratio.

Space for calculation

_____ : _____
0–2% : 6–8%

1

Page seventeen **[Turn over**

Marks | KU | PS

12. (*a*) School pupils each carried out an identical word processing task. The resulting level of muscle fatigue was measured on a scale from 1 (low) to 7 (severe).

The results for the 95 pupils tested are shown in the following bar chart.

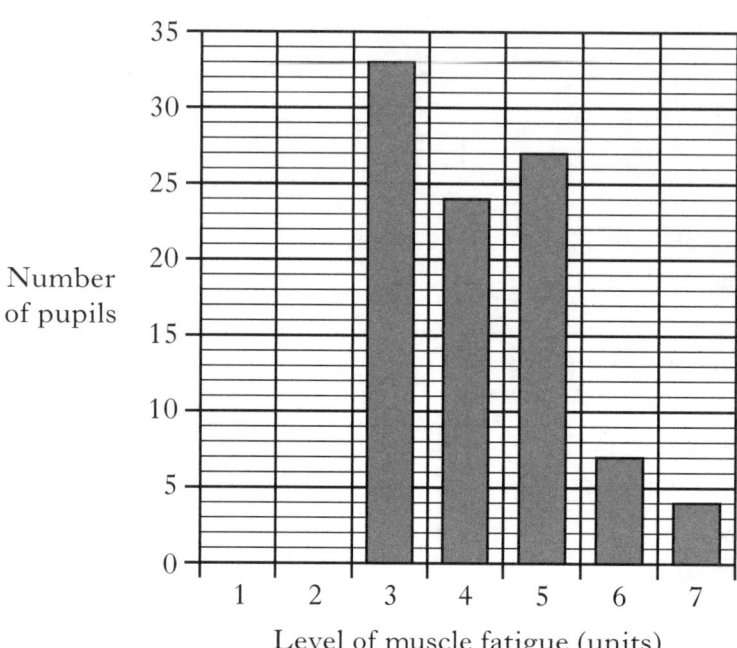

Number of pupils

Level of muscle fatigue (units)

(i) Medical experts using this scale classify any score of 5 or more as "requiring urgent investigation". What percentage of the pupils tested were in this category?

Space for calculation

_____ %

1

(ii) Give **two** conclusions which can be drawn from the results of this investigation.

1 _____

1

2 _____

1

(*b*) (i) What substance, produced by anaerobic respiration, causes muscle fatigue?

_____

1

(ii) Explain why ensuring an adequate blood supply to muscles reduces the risk of muscle fatigue.

_____

_____

1

Marks | KU | PS

13. The table below refers to egg production in the UK.

Living condition of hens	Eggs laid (percentage of total)
Living in cages	65
Living in barns	5
Free-range	30

(a) (i) Use the information from the table to complete the pie chart.

(An additional chart, if needed, will be found on *Page twenty-three*.)

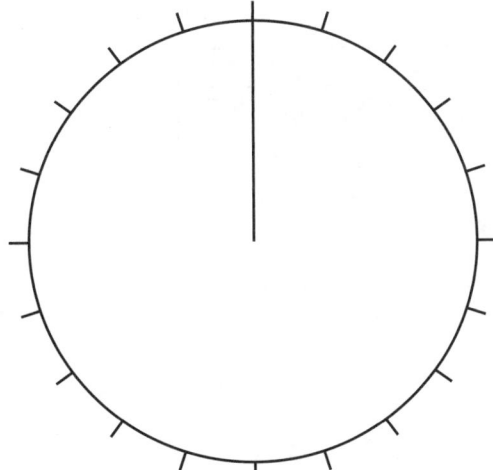

2

(ii) The total number of eggs laid per year in the UK is 30 million.

How many of these are laid by free-range hens?

Space for calculation

_____ eggs   1

(b) Modern varieties of hens can lay up to 300 eggs per year. Their ancestral wild varieties laid about 20 eggs per year.

(i) Calculate this increase in egg production as a percentage.

Space for calculation

_____ %   1

(ii) How has this improvement in egg production been achieved?

_____   1

Marks | KU | PS

14. Polydactyly is a condition which results in extra toes in mice. It is controlled by the dominant form of a gene (**N**). The normal phenotype is controlled by the recessive form (**n**).

The diagram below shows a cross between two mice of different genotypes.

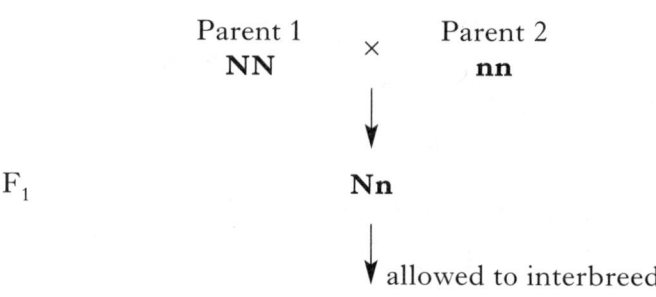

F$_1$ gametes	**N**	**n**
N		
n	**Nn**	

F$_2$

(a) (i) Complete the diagram above to show the possible genotypes of the F$_2$ generation.

1

(ii) Give the phenotypes of each of the following mice.

Parent 1 _____

Parent 2 _____

F$_1$ _____

2

(iii) What term is used to describe the type of variation shown by these phenotypes?

_____

1

(b) Why are the actual phenotype ratios in the F$_2$ generation often different from the predicted ones?

_____

_____

1

Marks | KU | PS

15. (*a*) Sucrose can be broken down into simple sugars using the enzyme invertase. The diagram below represents how this can be done commercially.

Sucrose solution is constantly being added and the products are constantly being removed.

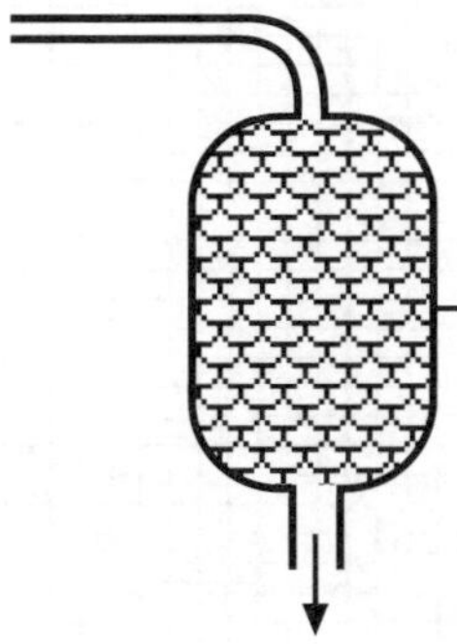

sucrose solution ⟶

reactor vessel
containing invertase

product rich in simple sugars

(i) What name is given to this type of process?

_____   1

(ii) Explain why the enzyme does not leave the reactor vessel along with the products.

_____   1

(*b*) (i) Genetic engineering techniques are used to produce enzymes which are used in biological washing powders. Which type of micro-organism is modified to produce the appropriate enzymes?

_____   1

(ii) What is transferred from one organism to another during genetic engineering?

_____   1

(*c*) During the brewing of beer, ingredients including yeast and malted barley are added to a fermentation vessel.

(i) What does the malted barley provide for fermentation which ungerminated barley does not?

_____   1

(ii) How does sterilising the fermentation vessel before the raw materials are added help to provide optimum conditions for the yeast?

_____   1

[Turn over for Question 16 on *Page twenty-two*

Marks KU | PS

16. The concentrations of lactic acid and lactose in a milk sample were measured every two hours for 100 hours. The results are shown in the graph below.

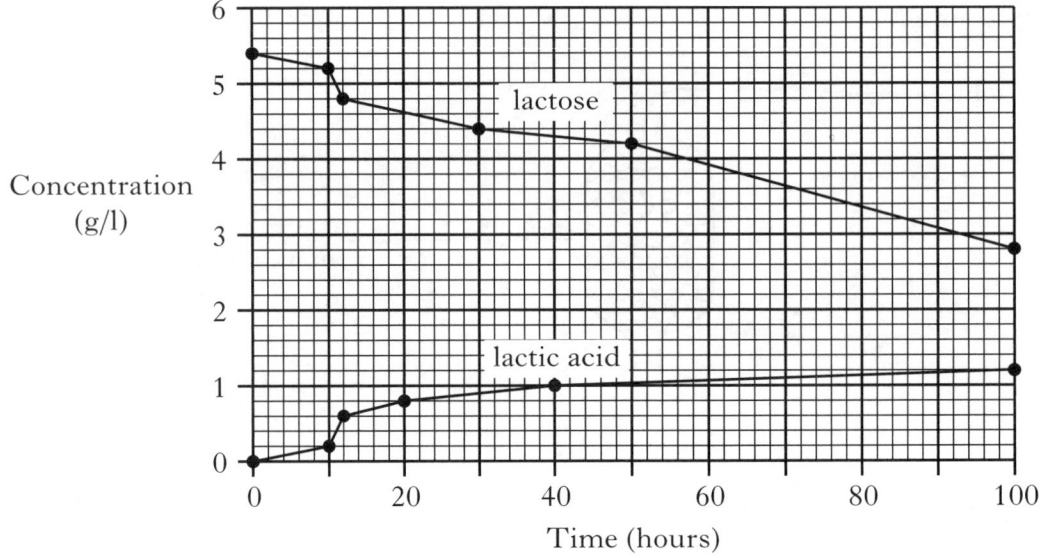

(*a*) (i) What evidence from the graph suggests that lactose is converted into lactic acid?

_____

_____ 1

(ii) What evidence from the graph supports the theory that lactose is being converted into compounds other than lactic acid?

_____

_____ 1

(*b*) Calculate the average hourly rate of lactose breakdown over the 100 hours of this investigation.

Space for calculation

_____ g/l/hour 1

[*END OF QUESTION PAPER*]

SPACE FOR ANSWERS
AND FOR ROUGH WORKING

ADDITIONAL GRID FOR QUESTION 6(a)

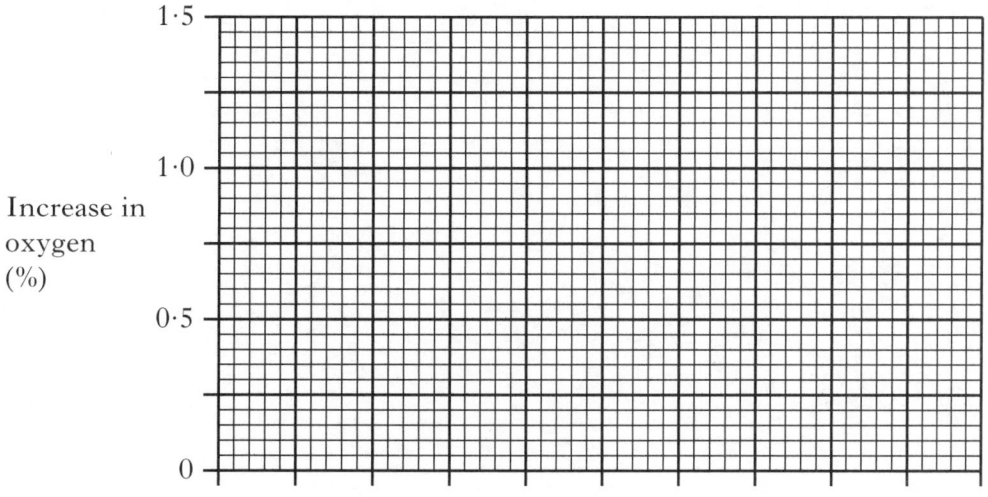

ADDITIONAL PIE CHART FOR QUESTION 13(a)(i)

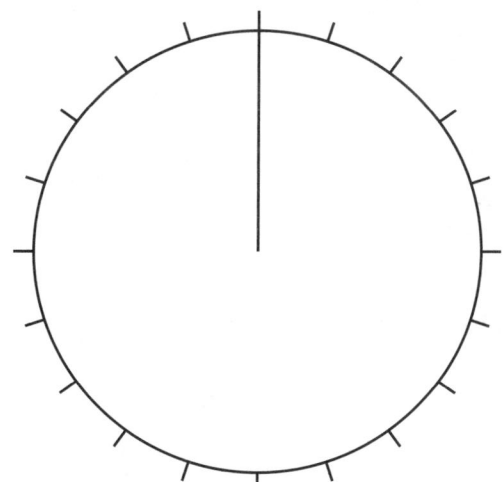

SPACE FOR ANSWERS
AND FOR ROUGH WORKING

[BLANK PAGE]

FOR OFFICIAL USE

C

KU   PS

Total Marks

0300/402

NATIONAL
QUALIFICATIONS
2011

MONDAY, 9 MAY
10.50 AM – 12.20 PM

**BIOLOGY
STANDARD GRADE**
Credit Level

Fill in these boxes and read what is printed below.

Full name of centre

Town

Forename(s)

Surname

Date of birth

Day   Month   Year   Scottish candidate number   Number of seat

1   All questions should be attempted.

2   The questions may be answered in any order but all answers are to be written in the spaces provided in this answer book, and must be written clearly and legibly in ink.

3   Rough work, if any should be necessary, as well as the fair copy, is to be written in this book. Additional spaces for answers and for rough work will be found at the end of the book. Rough work should be scored through when the fair copy has been written.

4   Before leaving the examination room you must give this book to the Invigilator. If you do not, you may lose all the marks for this paper.

Marks KU PS

1. Marsh marigold is a waterside plant which grows beside burns.

The abundance of marsh marigolds was estimated in five sampling areas beside a burn in the Scottish borders. Average values of three abiotic factors were also calculated for each area.

The results are shown in the table below.

Sample area	1	2	3	4	5
Abundance of marsh marigold	zero	high	high	medium	low
Average soil pH	5·6	6·7	7·1	6·5	6·4
Average soil nitrate concentration (ppm)	4	10	7	6	5
Average soil water content (units)	8	4	9	3	5

(*a*) Name **one** abiotic factor which does not affect the abundance of marsh marigolds.

_____   1

(*b*) The soil pH for each sampling area was measured using a pH meter with a probe which was pushed into the soil to obtain each reading.

(i) Identify a possible source of error in measuring a **named** abiotic factor and suggest how to minimise it.

Abiotic factor_____

Source of error_____

How to minimise it _____

_____   2

(ii) How was the measurement of the abiotic factors in this survey carried out to reduce the effect of atypical results?

_____   1

2.   The diagram below represents part of the nitrogen cycle.

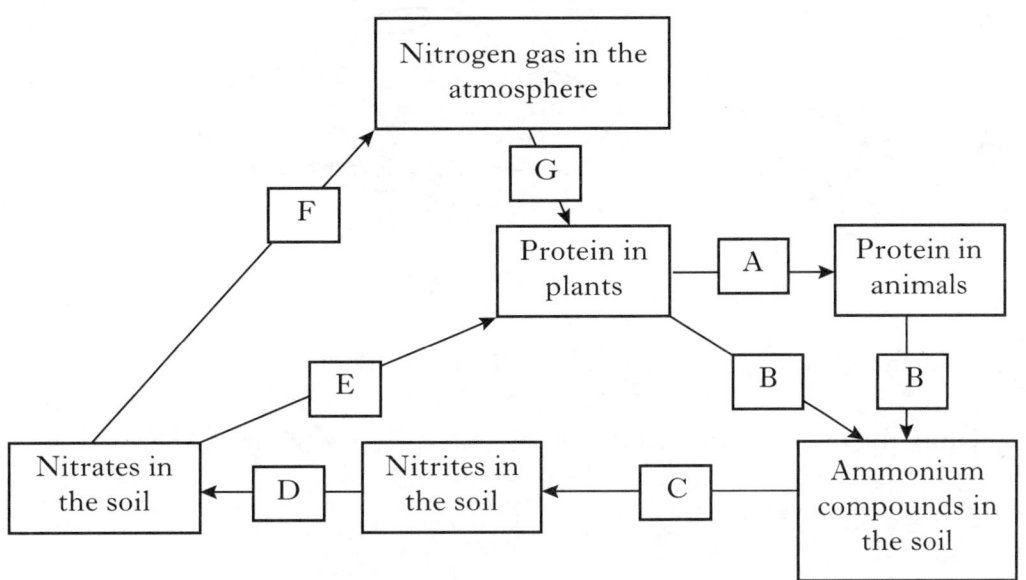

(a)   (i)   Use letters from the diagram to identify the following.

Each letter may be used once, more than once or not at all.

Decay of dead material   ☐

Nitrification   ☐☐

Nitrogen fixing   ☐   **2**

(ii)   Which type of organism is responsible for process D?

_____   **1**

(b)   In an investigation, wild rabbits were found to eat an average of 600 g of grass per day. This grass contains 450 g of water. The dry weight of the grass contains 20% protein.

Calculate how much protein a rabbit eats per day.

Space for calculation

_____ g   **1**

[Turn over

3. The diagrams below show two types of flower.

Diagram A　　　　　　Diagram B

(a) (i) Identify the insect pollinated flower, by putting a tick (✓) in the box.

Diagram A ☐

Diagram B ☐

1

(ii) Wind pollinated flowers produce larger quantities of pollen than those pollinated by insects. Explain why this is necessary.

_____

_____

1

(b) The table shows when some wind pollinated species start to produce pollen. Pollen production then continues for an average of five weeks.

Plant	Start of pollen production
Alder	February
Willow	March
Silver birch	April
Oak	April
Grasses	May

From the information given, why is May likely to be a particularly difficult month for people with pollen allergies?

_____

_____

1

Marks | KU | PS

3. (continued)

(c) (i) Sexual and asexual reproduction in plants have different advantages. For **each** advantage described in the table below, identify the method of reproduction involved.

Tick (✓) the correct box.

Advantage	Method of Reproduction	
	sexual	asexual
Variation exists amongst the offspring		
Germination is not required		
Desirable characteristics are maintained		
Seeds are produced which can be dispersed		

2

(ii) Underline the correct word in brackets to complete the sentence below.

A group of plants which are genetically identical is known as a

$$\left\{ \begin{array}{l} \text{clone} \\ \text{species} \\ \text{genotype} \end{array} \right\}.$$

1

[Turn over

Marks KU PS

4. An investigation was carried out into the effect of the concentration of a plant growth substance on shoot growth in seedlings. The length of each shoot was measured at the start of the investigation.

Seven solutions of the plant growth substance, each with a different concentration, were prepared. Ten seedlings were placed in each solution. A further ten seedlings were placed in distilled water.

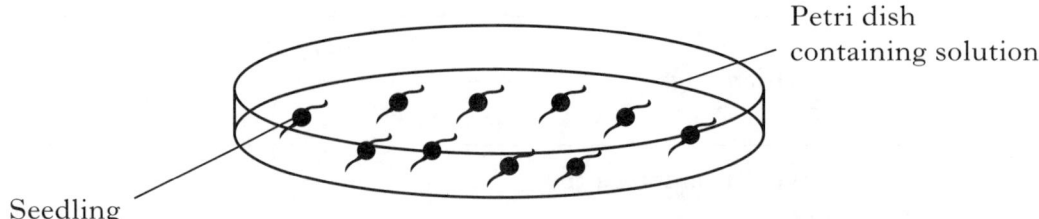

Petri dish containing solution

Seedling

After three days, the shoots were measured again and the results recorded in the table.

	Concentration of plant growth substance (ppm)							
	0	0·0001	0·001	0·01	0·1	1	5	10
Average length of shoot at start (mm)	5·0	5·0	5·0	5·0	5·0	5·0	5·0	5·0
Average length of shoot after treatment (mm)	10·0	10·0	10·4	12·3	17·0	11·6	9·6	6·3
Average increase in length of shoot (mm)	5·0	5·0	5·4	7·3	12·0	6·6	4·6	1·3

(a) Describe from 0·0001 ppm to 10 ppm the relationship between the concentration of plant growth substance and the average increase in shoot length.

_____

_____

_____   2

(b) Why was a set of seedlings grown in distilled water (0 ppm)?

_____

_____   1

5. The bar chart shows the average annual losses in yield caused by insects and disease in the production of three crops in Scotland.

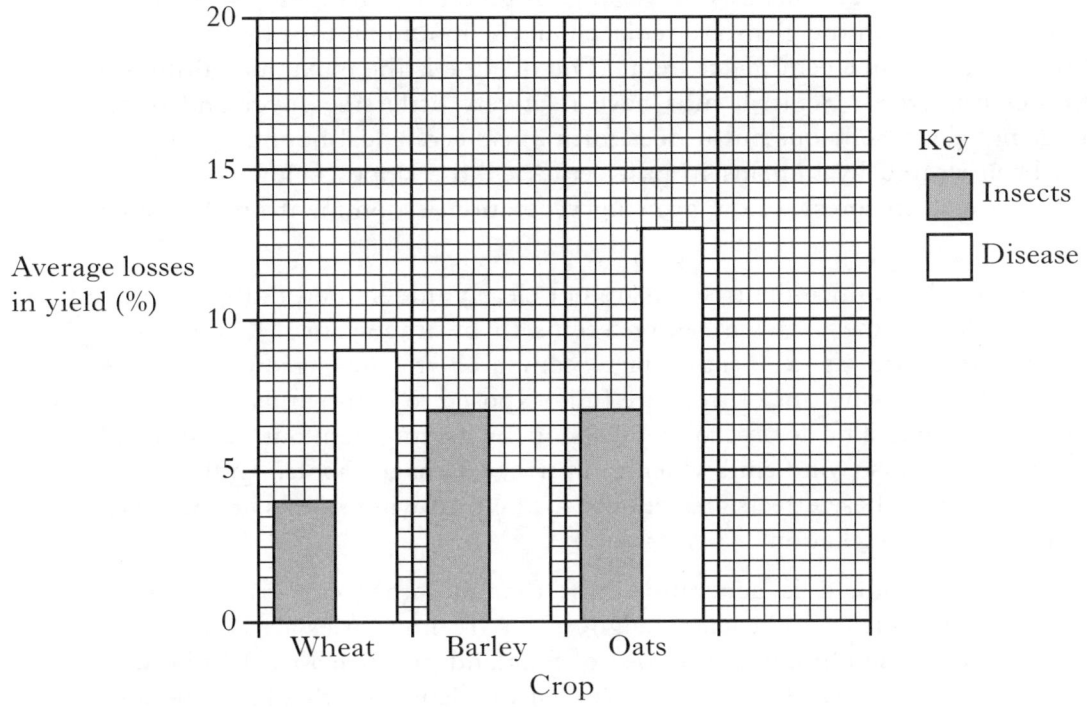

(a) (i) Which crop has the lowest combined percentage loss from these two causes?

_____

1

(ii) The total crop of oats harvested was 130 000 tonnes. Calculate the yield of oats which would have been produced if insects and disease had not affected the plants.

Space for calculation

_____ tonnes

1

(iii) Explain why it would **not** be a valid conclusion to say that disease caused more tonnes of oats to be lost than any of the other crops named.

_____

_____

1

(b) Oilseed rape is a common crop which has average annual losses of 9% to insects and 12% to disease.

Use this information to complete the bar chart by adding a label and bars in the space provided.

(An additional chart, if required, can be found on *Page twenty-six*.)

1

Marks | KU | PS

6. Read the passage below and use the information to answer the questions which follow. (Adapted from *Hostile Habitats*, Scottish Mountaineering Trust, 2006).

As you climb a mountain or hill, the vegetation gradually changes. In Scotland, trees and tall grasses in the glens are replaced on the mountain tops by lichens and dwarf mosses less than a centimetre high. The treeline is the maximum altitude at which trees can grow. Scottish hills have relatively little tree cover and so the treeline is not always obvious but it does form a real ecological boundary. If trees had not been cleared by humans in past centuries, the slopes below the treeline would be covered in forest. Low growing vegetation is dominant on the higher slopes.

The factors which produce the treeline are not clearly understood but the average temperature during the growing season seems to be important. Under colder conditions, trees are at a disadvantage compared to low growing, denser vegetation. The growing tips of trees are fully exposed to high winds which cause physical damage and slow down growth of shoots by drying them out. High winds in wet conditions cause wind chill which can further damage shoots. In the case of low growing plants, these effects are reduced as their growing shoots are protected by the surrounding vegetation.

The treeline in Scotland is generally lower than in other countries a similar distance from the equator. The exact height of the treeline varies across Scotland. The wet and windy conditions in the west of Scotland produce a treeline between 200 m and 450 m above sea level. In the east of Scotland, the treeline is between 500 m and 650 m above sea level. Other types of vegetation show similar effects, with mountain plants being found at lower levels on the west coast.

(*a*) Give **two** types of plants you might expect to find growing on mountain tops in Scotland.

1 _____

2 _____   1

(*b*) Most hills in Scotland do not have woodland present up to the potential treeline. Why is this?

_____

_____   1

(*c*) According to the passage, what factor might be important in determining how high up a hill trees can grow?

_____   1

(*d*) What **two** factors are needed to produce wind-chill?

1 _____

2 _____   1

DO NOT WRITE IN THIS MARGIN

Marks | KU | PS

6. (continued)

(*e*) The passage states that, "Low growing vegetation is dominant on the higher slopes". What advantage does this type of vegetation have which allows it to grow at higher altitudes than trees?

____

____ **1**

(*f*) In summer, red deer migrate to graze above the treeline. In which part of Scotland would they have to go higher to do this?

____ **1**

[Turn over

Marks   KU   PS

7. (a) Draw **one** line from each food component to the diagram which represents its basic structure.

Food component   Basic structure diagram

                                                        amino acids

carbohydrate

fat

                                                        glycerol

protein

2

(b) The following list contains structures associated with digestion.

Structures associated with digestion

A   gall bladder

B   large intestine

C   liver

D   pancreas

E   salivary glands

F   oesophagus

Use letters from the list to identify the structures which carry out the functions described below.

Each letter can be used once, more than once or not at all.

Function	*Structures*
Carry out peristalsis	
Produce amylase enzymes	
Produces digestive juices which are not enzymes	

2

Marks KU PS

7. (continued)

(c) The diagram below represents a structure found in the small intestine. The arrows show the direction of the flow of fluids through the structure.

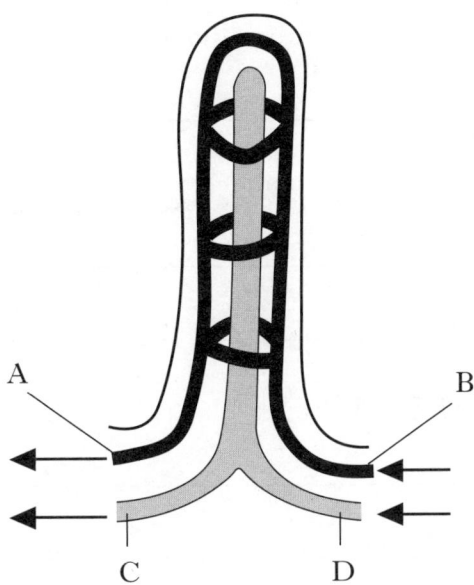

(i) What is the name of this structure?

_____

1

(ii) Which letter identifies the position of the fluid with the highest glucose content, after the absorption of digested food?

_____

1

(iii) Which letter identifies the position of the fluid with the highest fat content, after the absorption of digested food?

_____

1

[Turn over

Marks   KU | PS

8. (*a*) The process of diffusion is important to organisms.

From the list below, select a substance which is involved in diffusion and answer the questions which follow.

List

oxygen   glucose   carbon dioxide

Substance selected _____

(i) Explain why its diffusion is important.

_____

_____

(ii) Where does its diffusion take place?

_____

_____   2

(*b*) Cells from the same plant tissue were placed in three different liquids, left for 20 minutes and then examined using a microscope.

The following diagrams represent cells from each liquid.

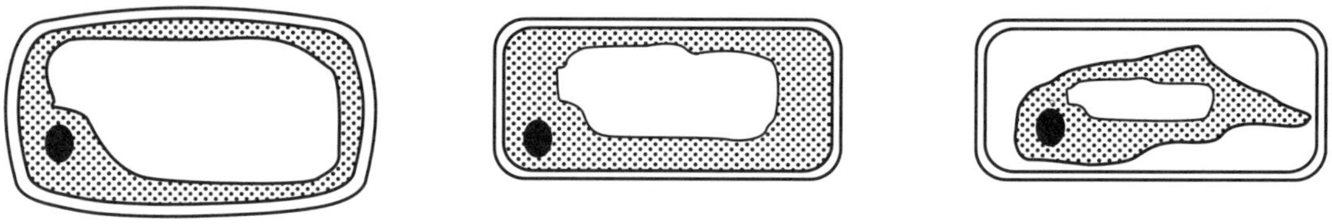

Cell A   Cell B   Cell C

Which cell is most likely to have been placed in pure water?

Give a reason for your answer.

Cell _____

Reason _____

_____

_____   1

8. (continued)

(c) The following is a description of the stages of mitosis.

| Stage 1 — Chromosomes become visible as pairs of chromatids |

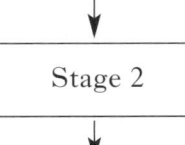

| Stage 3 — Pairs of chromatids attach to the spindle near the middle of the cell |

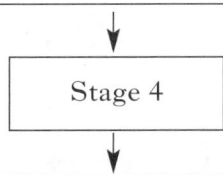

| Stage 5 — Daughter chromosomes gather at the ends of the cell |

| Stage 6 — The cytoplasm divides |

Describe stages 2 and 4 in the spaces below.

Stage 2 _____

_____

Stage 4 _____

_____ 2

(d) Daughter cells produced by mitosis each have the same chromosome complement as the original cell. Why is this important?

_____

_____ 1

[Turn over

Marks

9. An investigation was carried out into digestion of a protein.

   The protein was mixed with agar gel in a petri dish. Four holes were cut in the gel and a different enzyme was placed in each hole. The dish was left for two days. Where digestion of the protein had taken place, a clear area developed in the gel around the hole. The diameter of the clear area was measured. The experiment was carried out four times.

   The diagram below represents the appearance of one of the petri dishes after two days.

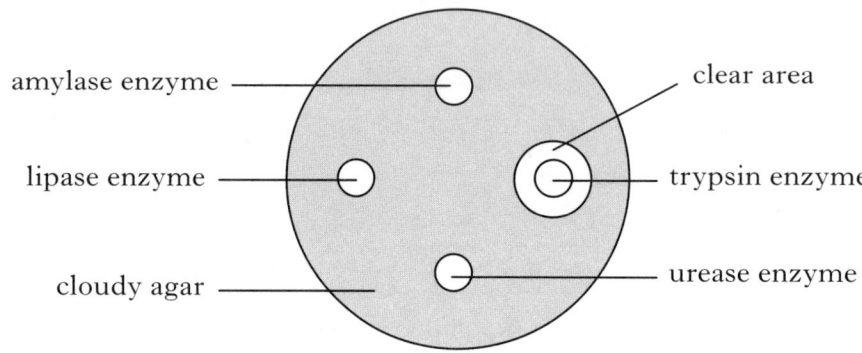

   (a) Explain why trypsin digested the protein but no other enzyme did.

   _____   1

   (b) The table below shows the results for each dish.

Petri dish	Diameter of clear area (mm) around trypsin enzyme
1	4·7
2	3·9
3	4·2
4	4·4
Average	

   Complete the table by calculating the average diameter of the clear area.

   Space for calculation

   1

   (c) Give **two** precautions, **not already mentioned**, that would have to be taken each time the experiment was carried out, to ensure validity of the results.

   1 _____

   2 _____   2

Marks

KU	PS

10. The following table shows the changes in the flow of blood through the capillaries in some body organs at rest and during exercise.

Body organs	Capillary blood flow (cm³/min)	
	At rest	*During exercise*
heart muscle	260	650
brain	760	760
skin	380	1200
intestine	1160	540

(a) Use the information from the table to complete the bar chart below.

(An additional chart can be found, if required, on *Page twenty-six*.)

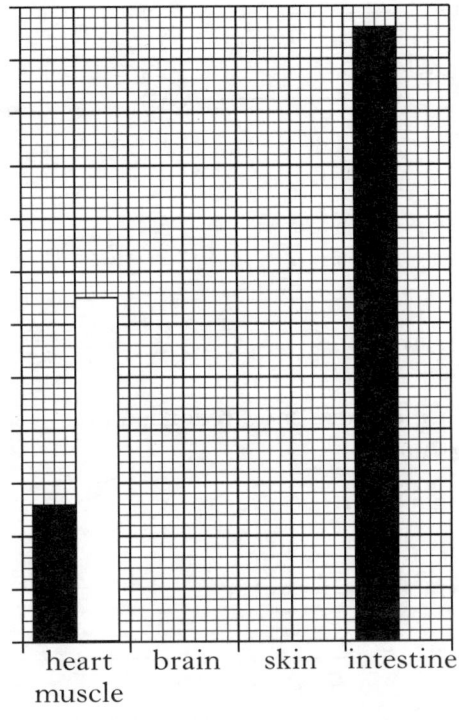

At rest

During exercise

heart muscle brain skin intestine

Body organs

2

(b) How does the capillary blood flow through the heart muscle at rest compare to that during exercise, expressed as a simple whole number ratio?

Space for calculation

_____ : _____

at rest   during exercise

1

(c) Suggest a reason for the decrease in blood flow to the intestine during exercise.

_____

_____

1

(d) Blood carries heat away from the muscles during exercise. What evidence from the table suggests that this heat is lost from the skin?

_____

1

Marks   KU | PS

11.   In an investigation into the energy content of a food, several samples were weighed before being burned. The heat energy given out was measured by noting the rise in temperature of the water.

Different methods were used by two different groups. The apparatus used by each group is shown below.

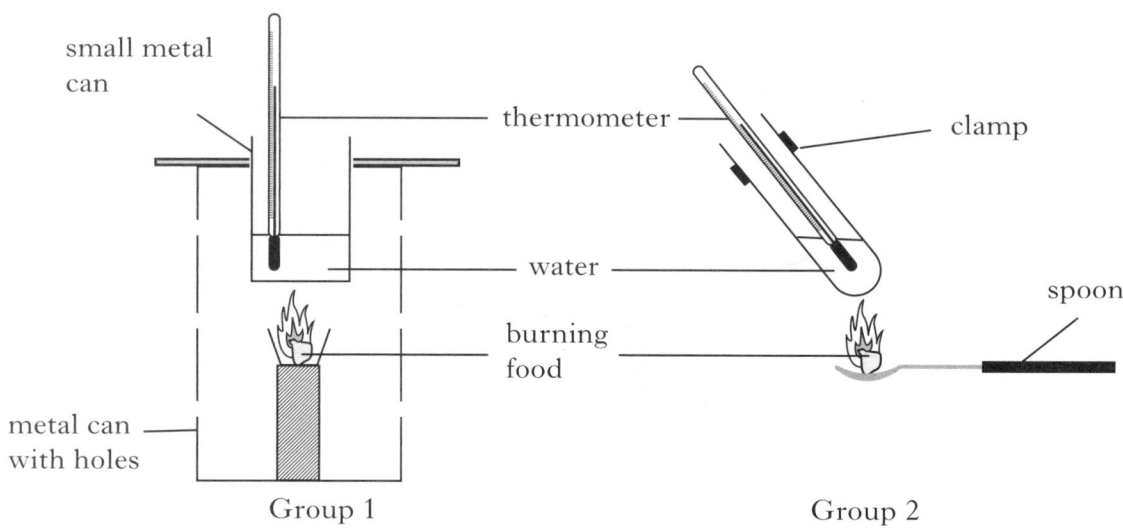

Group 1   Group 2

Both groups used the same mass of the same food.

(a)   Group 2 found that the rise in temperature they recorded was less in every case than the results obtained by group 1.

With reference to their method, suggest a reason for this.

_____

_____   1

(b)   Although they used different methods to investigate the energy content of the food, suggest a variable, **not already mentioned**, which both groups should have kept constant to allow a valid comparison.

_____   1

Marks

| | KU | PS |

12. The effect of practice on performance was investigated. The total score of a dart player with 3 darts was recorded for several attempts.

The scores are shown in the table below.

Attempt	Total score
1	32
2	36
3	43
4	58
5	65
6	64
7	65
8	64

(a) Describe the relationship between the number of attempts and performance.

_____

_____ 1

(b) How could this investigation have been made more reliable?

_____ 1

[Turn over

13. A pupil carried out an investigation into the effect of exercise on the body's heart rate. Using an exercise bike, he pedalled at different work rates for three minutes with a one minute rest between each exercise period.

During the exercise periods his heart rate was measured. The results are shown in the table.

Work rate (watts)	Heart rate (beats per minute)
0 (at rest)	80
60	104
80	110
120	128
140	140
160	158
200	180

(a) Use the results to complete a line graph of the pupil's heart rate over the range of work rates.

(An additional grid can be found, if required, on *Page twenty-seven.*)

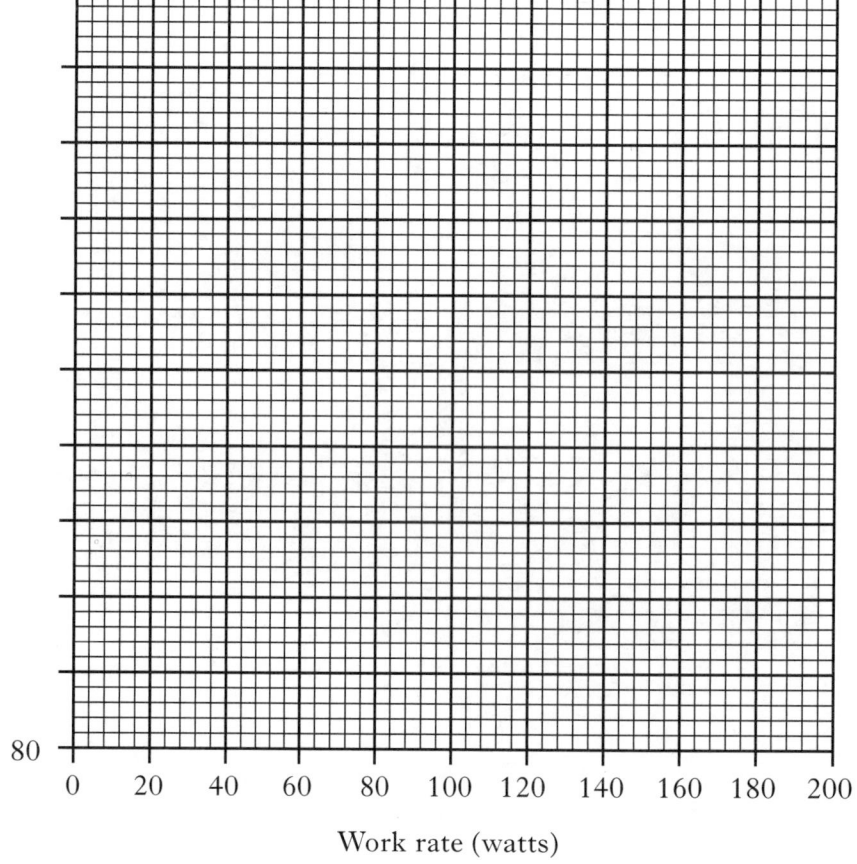

Work rate (watts)

2

DO NOT WRITE IN THIS MARGIN

Marks | KU | PS

13. (continued)

(*b*) Calculate the percentage increase in his heart rate from his resting state to a work rate of 200 watts.

Space for calculation

____________ % **1**

(*c*) Training through exercise improves the efficiency of the heart and other muscles. What other organs become more efficient as a result of training through exercise?

____ **1**

[Turn over

Marks KU PS

14. (*a*) An aircraft pilot must be able to sense accurately the movement of the aircraft when it is rolling, pitching or yawing, as shown below.

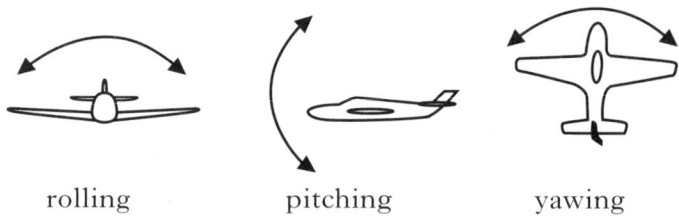

rolling   pitching   yawing

(i) Which structures in the pilot's inner ear can detect these movements?

_____   1

(ii) How does the arrangement of these structures make it possible to detect movement in these different directions?

_____   1

(*b*) The following diagram shows the field of vision of a cricket batsman viewed from above. The shaded section shows the area which can be seen by both eyes at the same time.

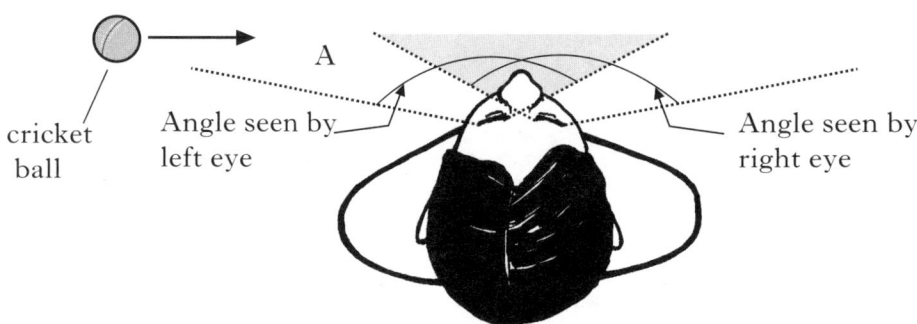

cricket
ball

Angle seen by
left eye

A

Angle seen by
right eye

What would be the advantage to the cricket batsman of turning his head towards the bowler so that a ball coming towards him appears in the shaded zone even though he could see it clearly in area A?

_____

_____   1

(*c*) The grid below shows structures related to the nervous system.

A	B	C	D
relay nerve cell	muscle	motor nerve cell	sensory nerve cell

Complete the sequence below, using letters from the grid, to show the order of the structures through which a nerve impulse travels in a reflex action.

stimulus → touch receptor → _____ → _____ → _____ → _____ → response   1

15. The difference between blue and green feather colour in budgerigars (budgies) is determined by a single gene. The allele for green (G) is dominant and the allele for blue (g) is recessive.

True-breeding blue males were allowed to breed with true-breeding green females. The offspring were allowed to interbreed to produce a second generation.

(a) Explain what is meant by the term "true-breeding", in terms of the alleles present.

_____   1

(b) Give the genotype(s) and phenotype(s) of the F_1 generation.

genotype(s) _____

phenotype(s) _____   1

(c) In 1974, a mutation occurred in a budgie which gave rise to one chick with a speckled pattern of wing feathers never before seen. Such birds are called "spangles". It is now 37 years since the hatching of the first chick, and the number of spangles now living is estimated to be 80 000 in a total population of 30 million captive budgies.

   (i) In which structures in the nucleus of a cell do mutations arise?

     _____   1

   (ii) Give an example of a factor which can influence the rate of mutation in an organism.

     _____   1

   (iii) Calculate the average yearly increase of spangles. Express your answer to the nearest whole number.

     Space for calculation

     _____   1

(d) Many varieties of budgies have been developed as a result of humans making a careful choice of which birds were allowed to breed over many generations. What name is given to this process?

_____   1

Marks KU PS

16. Antibiotics can be produced using immobilised enzymes.

Substrate in

Immobilised enzymes

Antibiotic out

(*a*) (i) What name is given to a process such as this where the product is collected without interruption for as long as the substrate is supplied?

_____   1

(ii) Give **two** advantages of using immobilised enzymes in this system.

1 _____

2 _____   2

(iii) This process was carried out at the optimum temperature for the enzyme. However, the antibiotic collected was not pure as it was mixed with some substrate.

Suggest a way to overcome this problem.

_____   1

(*b*) Several different antibiotics can be produced in this way. Why is it necessary to have a range of different antibiotics?

_____

_____   1

Marks | KU | PS

16. **(continued)**

(*c*) When antibiotics are prescribed, they need to be taken at **regular** intervals.

The pie chart below shows a 24 hour period, indicating sleep and waking hours.

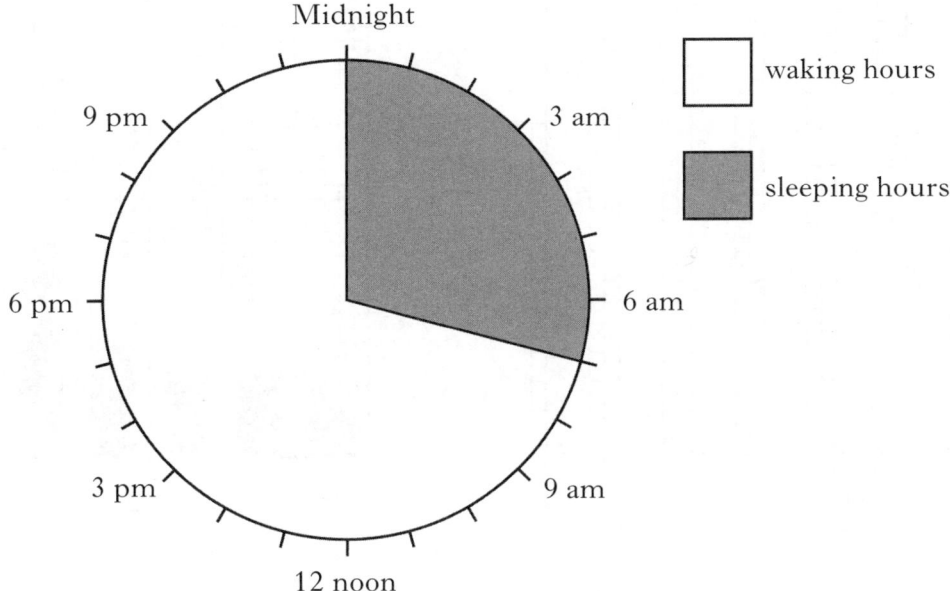

(i) If a patient took an antibiotic on wakening, and required two more that day, at what times should he take them to maintain a constant level in the body over 24 hours.

Space for calculation

1st _____   2nd _____   3rd _____   1

(ii) If the patient was given 3 grams of the antibiotic 3 times a day for a week, how much antibiotic was taken in total?

Space for calculation

_____ grams   1

[Turn over

17. The bar charts below show the mass of domestic waste produced and the percentage of that waste which was recycled in Scotland from 2001–2008.

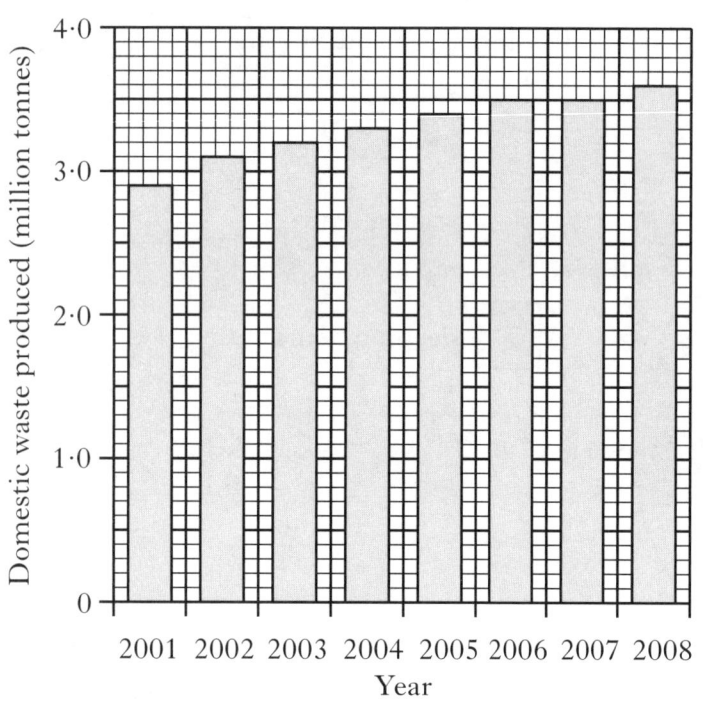

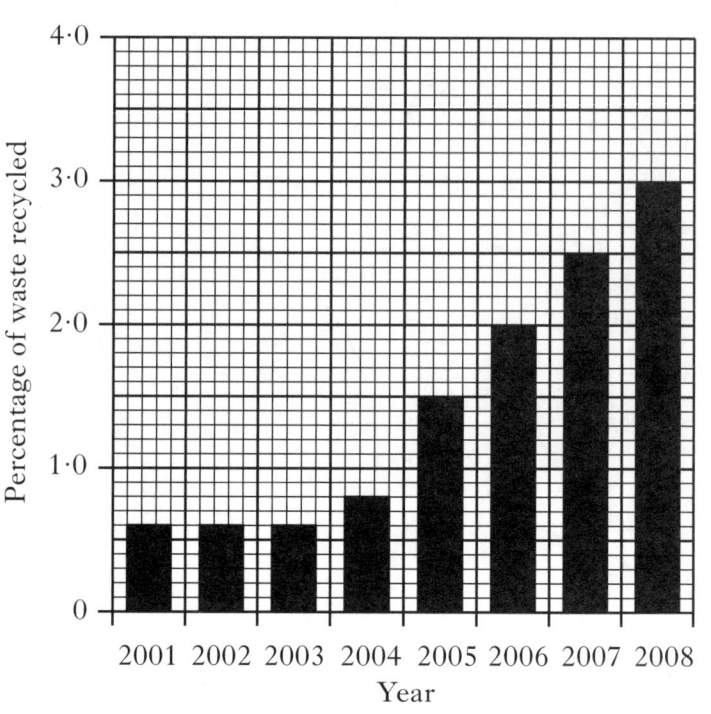

(a) Calculate the average yearly increase in production of domestic waste between 2001 and 2008.

Space for calculation

_____ million tonnes

1

(b) (i) Describe the percentage of domestic waste recycled in Scotland from 2001 to 2008.

_____

_____

2

(ii) How many million tonnes of domestic waste were recycled in 2006?

Space for calculation

_____ million tonnes

1

DO NOT WRITE IN THIS MARGIN

Marks KU PS

17. (continued)

(*c*) (i) Organic waste can be composted. This helps to recycle plant nutrients such as nitrates and minerals. Name **one** other element or compound, important for plant growth, which is recycled during decay processes such as composting.

____________________ **1**

(ii) After the manufacture of the compost is complete it may be treated with steam at 120 °C before it is sold. Explain the purpose of this treatment.

____

____ **1**

[*END OF QUESTION PAPER*]

SPACE FOR ANSWERS
AND FOR ROUGH WORKING

ADDITIONAL CHART FOR QUESTION 5(*b*)

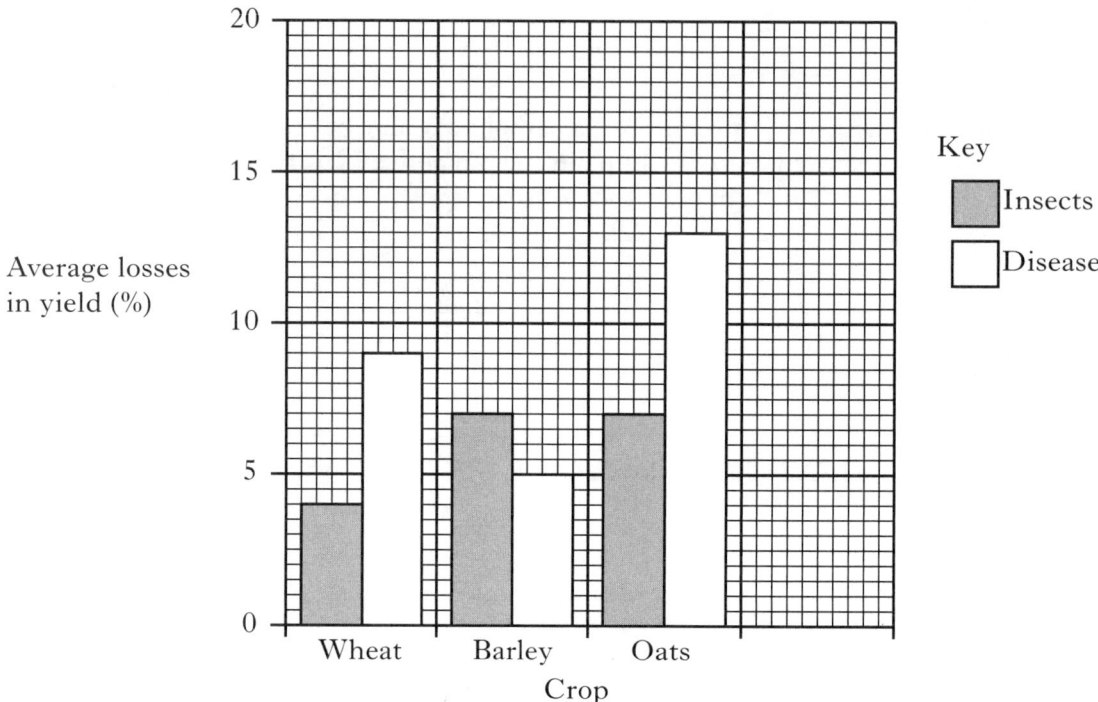

ADDITIONAL CHART FOR QUESTION 10(*a*)

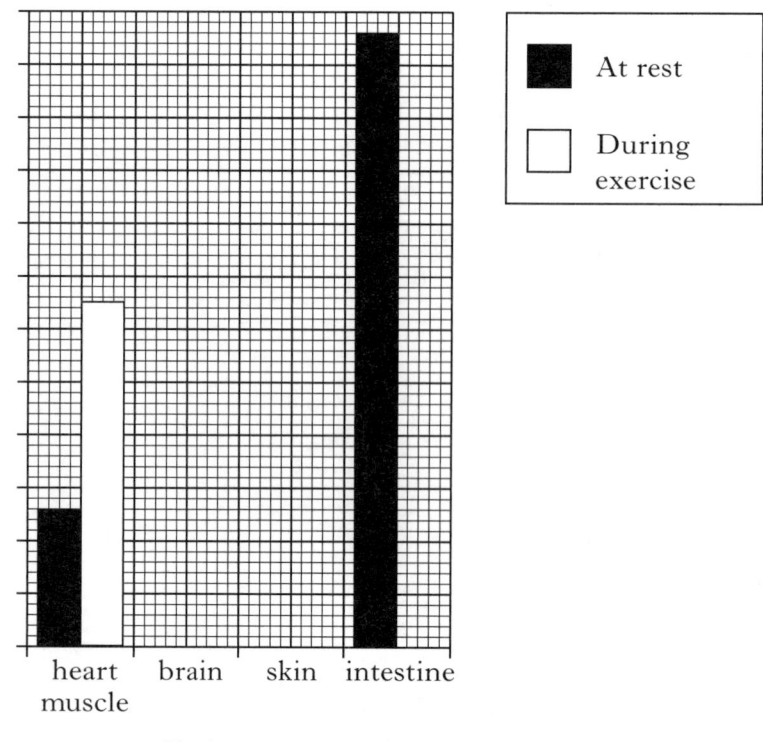

SPACE FOR ANSWERS
AND FOR ROUGH WORKING

ADDITIONAL GRID FOR QUESTION 13(*a*)

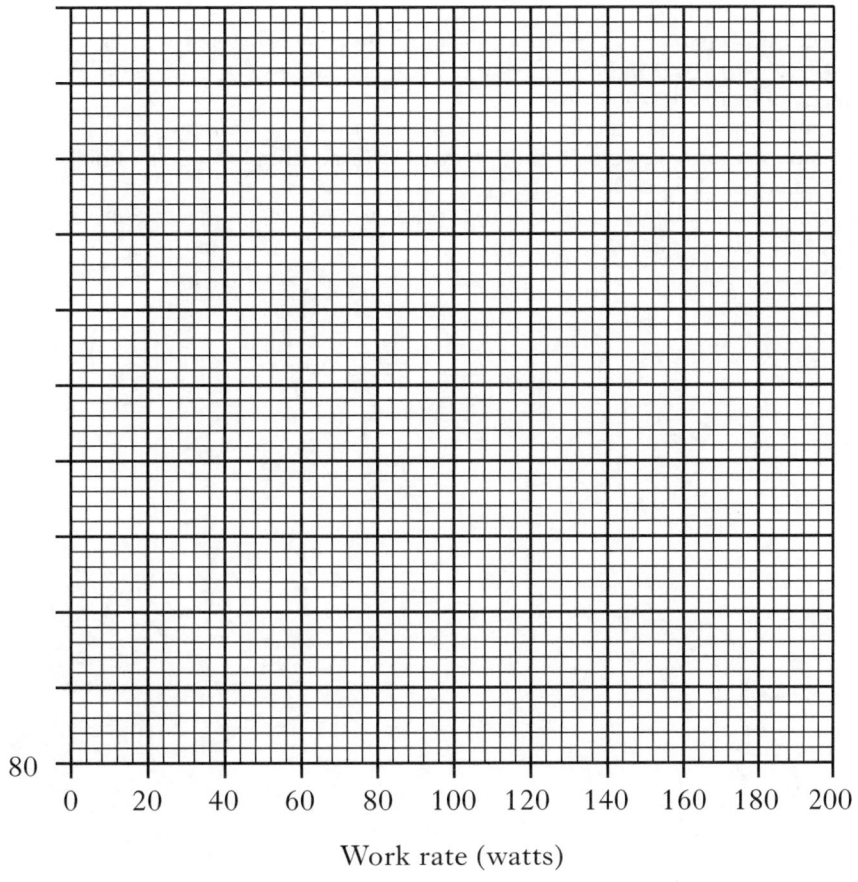

Work rate (watts)

Acknowledgements

Permission has been sought from all relevant copyright holders and Bright Red Publishing is grateful for the use of the following:

An article adapted from 'Invasion of the Chinese Mitten Crab', from Biological Sciences Review, Volume 15, Number 2 (November 2002) produced by permission of Philip Allan Updates (2007 page 16);

A photograph taken by Marilu Ormsbee (2008, page 18);

An extract adapted from 'GM Organisms' by John Pickrell, taken from www.newscientist.com (2008 page 22);

An extract from 'The Life of Birds' by David Attenborough, published by BBC Books (Random House). Reproduced by permission of Sir David Attenborough (2009 page 18);

An extract adapted from 'Coronary heart disease mortality among young adults in the US from 1980 through 2002. Concealed leveling of mortality rates' by Ford ES and Capewell S. Taken from J Am Coll Cardiol 2007; 50:2128–2132. Reproduced by permission of Dr ES Ford (2010 page 14);

An extract adapted from 'Hostile Habitats', produced by Scottish Mountaineering Trust (Publications) Ltd, 2006. Reproduced with permission (2011, page 8).

BIOLOGY CREDIT 2007

1. (a) (i) increases then: levels out / remains steady
   (ii) decrease in food / decrease in oxygen / build up of waste / build up of CO_2 / build up of lactic acid / decrease in pH
   (iii) any line showing continuous decrease in number of cells (line must reach the end of stage D or reach the X-axis)

   (b) energy value / protein content

2. (a)

Stage	Number
Absorption	5
Death	1
Nitrification	3 **or** 4
Decomposition	2

   (b) nitrate

   (c) nitrifying bacteria

3. (a) (i) used in respiration / for energy
      used to make cellulose / other organic compounds
   (ii) less light reaches leaves / some light is absorbed by the soot /
      pores / stomata are blocked and CO_2 uptake is reduced

   (b) (i) X in ovule
   (ii) allow male nucleus / male gamete / pollen nucleus to reach or fertilise ovule / female nucleus / female gamete

   (c) animals depend on plants for food / loss of food sources for animals
   animals depend on plants for shelter / loss of shelter for some animals

   (d) bright petals / scented flowers:
   dyes / scents / decorative use / attracts insects

   tough stem fibres:
   textiles / materials eg rope / high fibre diet / support for plant

   bitter seeds:
   food / medicine / stops seeds being eaten

   starchy root: food / food storage

4. (a)

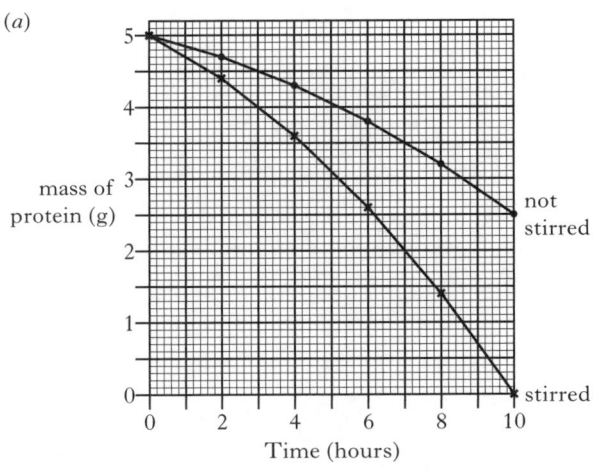

(b) protease / pepsin / trypsin

(c) products are soluble / have dissolved

(d) pH / enzyme concentration / type of protein / type of enzyme

(e) repeat investigation

(f) mixed enzyme and protein / increases contact between enzyme and protein

(g) contraction of muscles / contraction of the stomach wall / churning the food

5. (a)

	Name	Function
W	glomerulus	**filters blood/allows small particles through from the blood**
X	**Bowmans capsule**	collection of filtrate
Y	**blood capillary**	reabsorption
Z	collecting duct	**passes urine to middle of kidney/ureter**

(b) (i) liver
   blood
(ii) urea / glucose / amino acids / salts
   protein

(c) (i) 1045 : 95 : 1
(ii) 30·24

6. (a) 4

(b) dark
   high tide

(c) able to swim to find food at high tide / food present at high tide
   not seen by predators in darkness / food present when dark / avoids competition with animals which feed during the day

7. (a) (i) osmosis
   (ii) water moved out of cell from a high to low water concentration

(b) *Any one from:*
   • oxygen
   • glucose
   • amino acid

(c) 0·4
   0·2

8. (a) (i) *Any two from:*
   • volume of water
   • distance between burning food and test tube
   • size or type of test tube
   • starting temperature of water
   • complete burning of food
   (ii) *Any one from:*
   • heat energy escaping
   • energy lost as light
   • incomplete combustion
   • residual heat in tube or food or needle
   (iii) 6·3

(b) *Any one from:*
- growth
- cell division
- chemical reactions
- movement
- absorbing materials

(c) fat / oil

9. (a) Part X = ligament
    holds bones together / holds joint together / attaches bone to bone
    Part Y = cartilage
    protects bones / cushions joint / cushions bone / shock absorber / reduces friction

    (b) (i)   synovial membrane
               lubricates joint / reduces friction / nourishes cartilage
         (ii) 1. Fluid pink   Blood leakage
                  Fluid not pink   go to 2
               2. Low viscosity   **go to 3**
                  High vicosity   **normal**
               3. **High cloudiness** Infection
                  Slight cloudiness **Inflammation**

10. (a) claws covered in fine hairs

    (b) (i)   Sea / sea and estuaries
         (ii) estuaries
         (iii) freshwater

    (c) ships' ballast water

    (d) Habitat - collapse of banks / silting of rivers / destroying river banks
        Community - eating native species / competing with native species / competing with crayfish

    (e) herbivorous when young, becoming omnivorous

    (f) $2\frac{1}{2}$ / 2·5

11. (a) anaerobic

    (b) (i)   1·8
         (ii) 30·5
         (iii) 1. muscle fatigue
               2. 25
         (iv) heart
               lungs

12. (a) (i)   rr
               Rr
               Rr
         (ii) 1 in 2 / 1:1 / equal / even / 50% / 50-50
         (iii) Fred
         (iv) Jim / Margaret

    (b) fertilisation is random / fertilisation involves chance / numbers are too small

    (c) allele / alleles

13. (a) (i)   B
         (ii) C
         (iii) E

    (b) because heat is produced (during fermentation)

    (c) (i)   batch (process) / batch processing
         (ii) heat to 120°C / clean with disinfectant / clean with bleach

    (d) (i)   increases to maximum at 3 days / increases for 3 days / increases to optimum at 3 days then decreases
         (ii) lack of nutrients / yeast has used all the maltose / accumulation of waste / accumulation of alcohol

         (iii) to convert starch to sugars or maltose / because yeast can't use starch

14. (a) E

    (b) B + D

    (c) *Any one from:*
    - no one antibiotic is effective against all bacteria
    - bacteria may be resistant to some antibiotics
    - new bacterial strains appear
    - people can be allergic to some antibiotics
    - different antibiotics kill different bacteria

BIOLOGY CREDIT 2008

1. (a) (i) *Any two from:*
- size or area of sheet
- force or duration of shaking
- number of times shaken
- height of branches shaken
- size or age of tree
- position or aspect of tree
- time of year

(ii) *Any one from:*
- some fly away or walk away or escape
- some are not dislodged or cling too tightly
- some live on other parts of tree
- upper branches not sampled

(iii) *Any one from:*
- *Technique*: measuring light intensity / using light meter
- *Source of error*: casting shadow / pointing meter away from light source
- *Minimising*: avoid standing over meter

- *Technique*: measuring soil pH / using pH meter
- *Source of error*: probe damp / dirty
- *Minimising*: wipe probe

- *Technique*: measuring temperature using thermometer
- *Source of error*: handling thermometer bulb / not allowing reading to settle
- *Minimising*: don't handle bulb / allow to settle

- *Technique*: measuring soil moisture / using moisture meter
- *Source of error*: probe damp / dirty
- *Minimising*: wipe probe

(b) (i) **1** 2000
2 65

(ii) *Either:* as the light intensity increases, the ground cover increases
Or: as the light intensity decreases, the ground cover decreases

(c) the more light, the more plant growth / photosynthesis

2. (a) (i) **1** *Either:*
increase: less competition for food / more fruits and seeds or food to eat *or:*
decrease: more dormice would be eaten by owls *or:*
stay the same: both the above explanations given together with an indication that they would cancel each other
2 *Either:*
decrease: fewer sources of food
Or:
stay the same: eat more dormice or voles or weasels

(ii) *Any one from:*
- tree (bark) → woodlouse → weasel → fox
- tree (bark) → woodlouse → weasel → owl
- fruits and seeds → vole → weasel → fox
- fruits and seeds → vole → weasel → owl

(b) blue whale

3. (a) (i) C
(ii) F and G
(iii) D

(b) villi

(c) amino acids / protein

4. (a) keep trachea or windpipe or bronchi open / to prevent collapse or crushing of trachea etc / to support trachea etc

(b) (i) mucus
(ii) by the beating / sweeping / movement of cilia

(c) (i) *Any one from:*
- provides a large surface area
- have thin walls / surface
- have moist surfaces

(ii) red blood cells
(iii) forms oxyhaemoglobin

5. (a) (i) sieve plate
companion cell
(ii) transport of sugars or glucose or food or products of photosynthesis

(b) (i) epidermis
(ii) guard cells

(c)

	Only photosynthesis	Only respiration	Both	Neither
1		✓		
2			✓	
3				✓
4			✓	

6. (a) (i) 160
(ii) 1996 5:2
2005 5:1
(iii) The number of patients waiting for a transplant increased each year.

(b) Advantage:
Any one from:
- no dialysis
- doesn't need use of a machine
- no frequent or long hospital sessions
- can lead reasonably normal life

Disadvantage:
Any one from:
- risk of rejection
- need for medication
- risks involved with operation

7. (a) *Either:* as the boron concentration increases the greater the growth or length.
Or: as the boron concentration decreases the less the growth or length.

(b) 8

(c) to prevent any other source of boron affecting the results / other water may contain some boron / so the exact boron concentration was known / so no other minerals were present

(d) their food may contain boron / affect growth

8. (a) *Any two from:*
- size or volume of beaker
- volume of water
- depth of funnel in water
- diameter of glass tube
- diameter or width of funnel or area of membrane
- temperature
- membrane material (thickness / type etc)

(b) water entered the tube or funnel / because the water concentration was higher outside the funnel than in

(c) 31·5 mm

Moves 4·5 mm for each 0·5% increase in concentration.

9. (a) **Stage 2:** *Either:* chromosomes or chromatid pairs become attached to the spindle fibres
Or: chromosomes or chromatid pairs line up at the cell equator / middle of cell
Stage 5: *Either:* cell divided into two daughter cells
Or: cytoplasm divides
Or: new cell wall divides the cell

(b) • so there is no loss of information
• so cells have all necessary information

10. (a) *Any one from:*
• Proctor barley has bent stem / Rika barley has straight stem
• Proctor barley has awns which spread out from grains / Rika barley has awns close to grains
• Proctor barley is a lighter colour

(b) (i) to convert starch into sugars or maltose for the yeast / yeast cannot use starch
(ii) prevent oxygen entering fermenter / making sure conditions are anaerobic

11. (a) (i) alleles
(ii) The parents have **different** phenotypes and **different** genotypes.
(iii) Dd

(b) (i) 3:1
(ii) *Any one from:*
• random effects of fertilisation
• fertilisation involves chance
• sample size is too small
(iii) Tall P, Dwarf P

12. (a) fermentation / anaerobic respiration

(b) As the temperature increased up to 40°C the volume increased. As the temperature increased further, the volume decreased

(c) *Prediction* - any value below 5
Explanation - yeast cells killed / yeast enzymes denatured

(d) flasks with 100 cm³ of glucose solution and 50 cm³ boiled and cooled or dead yeast suspension kept at same range of temperatures

(e)

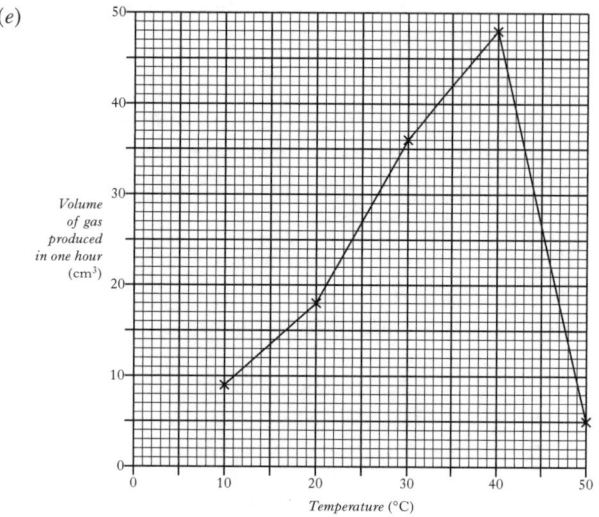

13. (a) transfer of useful genes from unrelated species

(b) enhanced protein content / contain more protein

(c) (i) allows use of weedkillers to remove weeds without harming crop
(ii) the modified plant becomes a pest which is difficult to control

(d) *Any one from:*
• modified crops may produce toxins
• antibiotic resistance genes may transfer to gut bacteria
• gut bacteria may become antibiotic resistant
• medicine production genes may transfer to food crops
• food crops may produce medicines

14. (a) (i) 45
(ii) 20–30 hours
(iii) 17·5 g / 100 cm³
(iv) 5 hours
(v) **1** glucose is being used
2 no hormone is produced

(b) fat / oil

15. (a) (source of) food / energy

(b) aerobic

(c) a large variety of micro-organisms

(d)

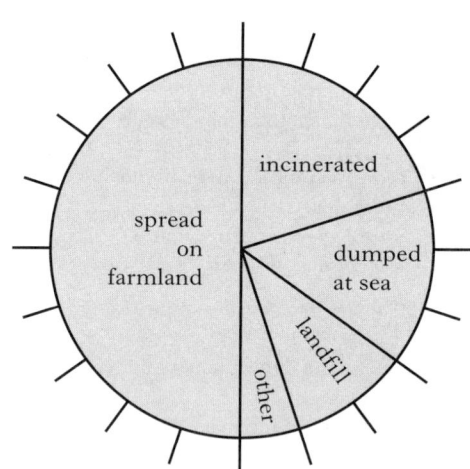

BIOLOGY CREDIT 2009

1. (a) (i) (A) → B   remained steady / does not change
         B → D   increased /
         D →   remained steady / does not change

   (ii) no lack of food / territory
        lack of or no predators / lack of or no disease

   (b) (i) increase (in population)
        less or no competition for <u>food</u> / more <u>food</u> / <u>grass</u> for them
        or
        (population) stays the same
        more <u>food</u> for wallabies but more wallabies are eaten

   (ii) decrease (in population)
        fewer prey / less <u>food</u>
        or
        (population) stays the same
        additional wallabies are eaten to make up for lack of rabbits

2. (a)

Coal burning
1 + 3
— waste can cause high levels of acid rain
— waste must be sealed before it is stored

Nuclear
2 + 4
— high volume of greenhouse gas production
— waste is dangerous for hundreds of years

   (b) (i)

pH or PH	oxygen (O_2) Saturation (%)	Suspended solids (mg/1)
7·9	91·5	4·0/4
7·7	65	5·6
8·0	94	6·0

   (ii) Mains (Burn)

3. (a) anther – produces light <u>pollen</u> grains / produces lots of <u>pollen</u> / hangs outside (flower) so <u>pollen</u> is (easily) blown away

   stigma – large surface area or feathery to catch / trap <u>pollen</u> / hangs outside (flower) to trap <u>pollen</u> / exposed to wind blown <u>pollen</u>

   (b) (i) 4

   (ii) sufferers not allergic to all pollen / different sufferers are allergic to different plants / pollen

   (iii) 75600

   (c) germination of pollen grain / growth of pollen tube / passage of pollen nucleus or male gamete to ovule or female gamete

   (d)

plant	sycamore / birch / ash	dandelion / thistle / willow / clematis
description	winged seeds / large surface area for wind to move them / helicopter like	seeds with fine hairs / light and feathery / umbrella like

   (e) 1 and 4

4. (a)

1.		(go to) 4
2.	**Black <u>wing tip</u>**	(go to) 3
3.		**Wood White**
4.	**White <u>wing tip</u>**	**Red Admiral**

   (b) (i) butterflies appeared earlier (in 2006)

   (ii) indicator species

5. (a) (i) 80

   (ii) Humans have greater survival rates because of greater parental care / internal development / lower predation

   (b) carbon dioxide / urea

6. (a) (i) osmosis

   (ii) fresh (water)

   (b) (i) burst / swell up and burst / goes turgid and bursts

   (ii) water is moving against a concentration gradient / from a lower to a higher water concentration / there is a higher water concentration in the vacuole than in the cytoplasm

7. (a) increases
              decreases
              decrease

   (b) (i) A

   (ii) E

   (c) (i) protein

   (ii) glucose

   (d) 178·2

8. (a) (i)

Stage A   Stage B   Stage C   Spindle

   (ii) Chromosomes / chromatids reach poles / opposite ends of cell / spindle disappears / formation of nuclei / formation of nuclear membrane

   (b) 12·5

   (c) so no loss of information or instructions / so daughter cells have the same information (as the mother cell) / No loss of genes

9. (a)

Name	Name
Cartilage	Synovial fluid
Function	Function
cushions or protects bones / shock absorber / absorbs impact / reduces friction / allows smooth movement / stop bones rubbing together	lubricates (joint) / reduces friction / allows smooth movement

   (b) To pass / transmit force or contraction of muscle to bone / to make bone move / to make limb move

   To pass / transmit movement of muscle to bone

   If they were elastic, force or contraction would not be passed to bone and limb / would not move

10. (a) 1 4 6 7

    (b) (i) 800

    (ii) anaerobic respiration

    (iii)

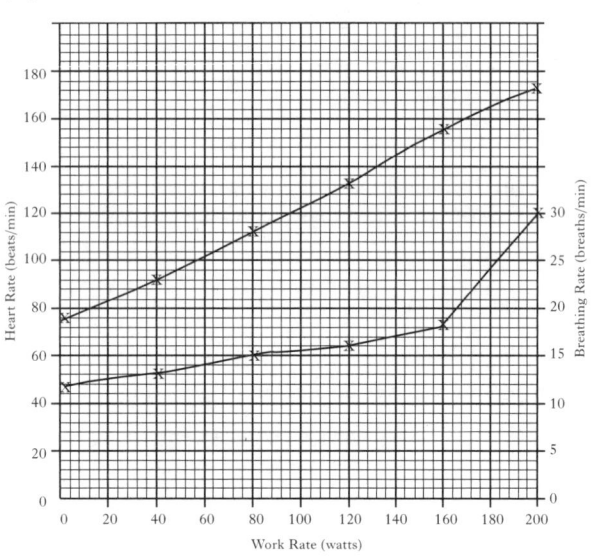

    (iv) As work rate increases, breathing rate and heart rate increase

11. (a) grit

    (b) oxygen
    to allow aerobic respiration of bacteria / micro-organisms /
    to provide aerobic conditions for micro-organisms /
    to allow <u>complete</u> breakdown of sewage / organic matter

    (c) A variety of micro-organisms is needed to breakdown the <u>range of organic matter</u> / the <u>different substances</u>

    (d) heavy rain / flooding

12. (a) to feed / to get food

    (b) 120

    (c) puts it on the top of his feet / with his upwardly turned toes

    (d)

1	2	3	4	5
Mar	Apr	May	Jul	Nov

    (e) 3

13. (a) (i)

    (ii) So only light from the cloth was recorded / So natural / outside light did not affect results.

    (b) (i) 30

    (ii) £31·20

    (iii) enzymes

    (iv) enzymes are denatured

14. (a)

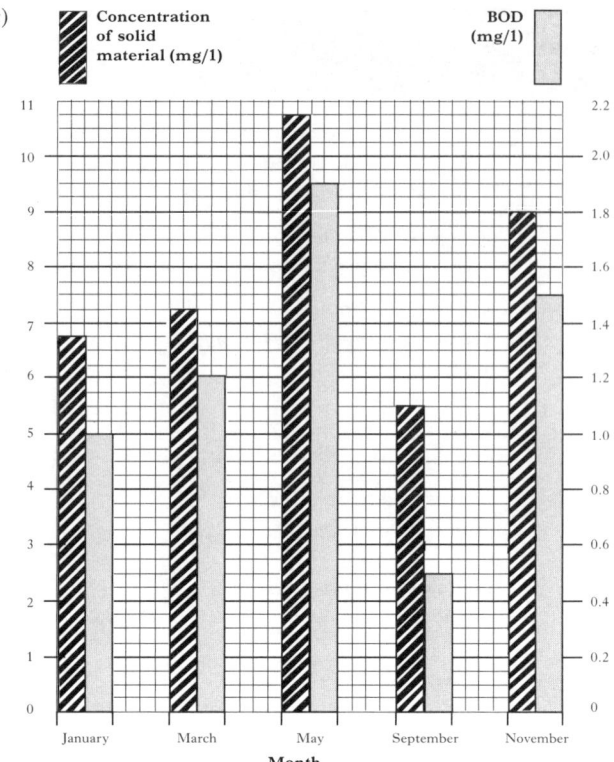

    (b) As the concentration of solid material decreases, the BOD decreases

    (c) 1·75mg / 1

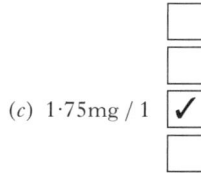

15. (a) (i) 3 : 1

    (ii) 12

    (iii) 3 : 2

    (iv) Random or chance effect of fertilisation / sample size too small

    (b) allele

    (c) (individuals) can be placed into distinct groups / (variation) does not show a range of values between a maximum and minimum / (variation) shows clear cut differences

BIOLOGY CREDIT 2010

1. (*a*) (i) 4

  (ii) 2

  (iii) Bigger sample used / Used more traps

  (iv) *Precaution*   *Reason* (must be appropriate to precaution)

  Pitfall trap

  Rim at soil level   So animals **or** invertebrates **or** insects can fall in

  Cover trap   So predators can't eat trapped animals / To keep rain out

  Drainage holes   So trapped animals don't drown

  Put alcohol in trap To preserve trapped animals / So trapped animals don't eat each other

(*b*) (i)   4
                   Spider
      No shell   *Earthworm*

  (ii) Spots on body
      Fewer than 12 legs
      Legs

2. (*a*) *Fossil fuel*   Limited supply / Finite / Greenhouse gas production / CO_2 production / SO_2 production / Causes acid rain / Causes global warming / Smoke causes asthma

  Nuclear fuel   Danger of radiation leaks / Waste is radioactive / Waste needs stored for a long time / Description of how waste must be stored

(*b*) (i) 1 There was more food / energy for the micro-organisms

       2 Micro-organisms use more oxygen / More micro-organisms using oxygen − not enough for fish

  (ii) Organisms which give information about the environment / pollution
      Organisms which live in specific conditions

3. (*a*) (i) 27

  (ii) volume or amount of water or moisture (or equivalent) / type of grass seed / pH

  (iii) 20 seeds used / large number of seeds used

(*b*) As temperature increases up to an optimum **or** 27°C, percentage germination increases
     As temperature increases further, percentage germination decreases

4. (*a*) (i) 2 and 6

  (ii) 46

  (iii) 2

  (iv) Length of all the roots added together **or** Total length of roots
       and
       Divide total by the number of roots

(*b*) Can be sure of their characteristics / Show same features **or** characteristics **or** good points as parent / All will be as successful as parent / Avoids vulnerable early stage of growth / Quicker

(*c*) Clone

5. (*a*) (i) 1 800
              2 0.05

  (ii) Fewer eggs **or** young surviving / More chance of eggs not being fertilised / More chance of eggs **or** young being eaten / Eggs are less well protected
      (answer needs a comparison)

  (iii) No external water for sperm to swim / Sperm need fluid to swim

(*b*) placenta
     exchange of materials between mother and fetus **or** embryo **or** baby / passes food **or** oxygen **or** nutrients from mother to fetus / passes waste **or** urea **or** CO_2 from fetus to mother

6. (*a*)

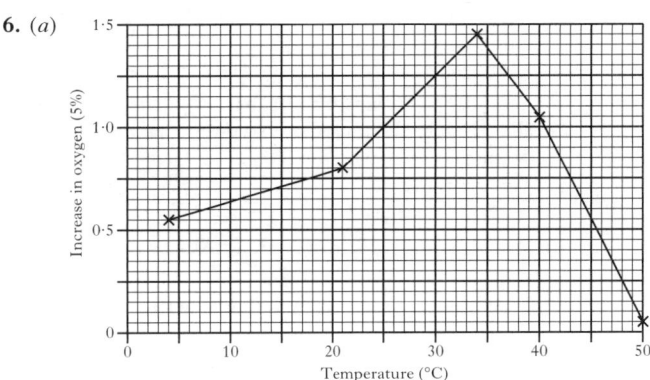

(*b*) 34

(*c*) To avoid reaction starting at wrong temperature / So it was at the correct temperature at the start of the reaction / Because reaction would start as soon as the catalase was added

(*d*) Other enzymes do not break down hydrogen peroxide / Enzymes are specific / Enzymes only work on one substrate / Other enzymes have different substrates

(*e*) 29 : 21 : 1

7. (*a*) 0.9
     No gain or loss of water at this concentration / No osmosis at this concentration / No change to cells at this concentration / They look like the untreated cells

(*b*) Cells have shrunk **or** become crenated **or** crinkled up **or** shrivelled up

     Water has moved out of cell by osmosis / Water has moved out of cell to a lower water concentration / Water has moved out of cell down a concentration gradient

8. (*a*) (i) One muscle moves joint in one direction **and** second muscle needed to move joint in opposite direction **or** One muscle bends joint **and** second muscle needed to straighten joint
      Muscles only work in one direction **and** Two muscles needed for full movement

  (ii) Tendons are inelastic / do not stretch

(*b*) cartilage
     synovial fluid

9. (*a*) *Any three from:*
     obesity / diabetes / high blood pressure / smoking / lack of exercise / hardening of the arteries

(*b*)

2.9	2.6	4.4
6.2	2.3	0

(c) Changes (or example) that lead to heart disease occur at an early age / To reduce risk of heart disease later

(d) Respiration

10. (a) (i) 7.5
  (ii) Fat

(b) carbon   hydrogen   oxygen / C H O

11. (a) decreases
    increases

(b) (i) 5

  (ii) 0.64

  (iii) 1 : 9

12. (a) (i) 40

  (ii) *Any two from:*
    Word processing causes muscle fatigue **or** All pupils showed muscle fatigue **or** No pupils had no fatigue **or** No pupils had very low fatigue

(b) (i) lactic acid

  (ii) Reduces anaerobic respiration / Allows more aerobic respiration
    Increased oxygen supply reduces lactic acid production

13. (a) (i)

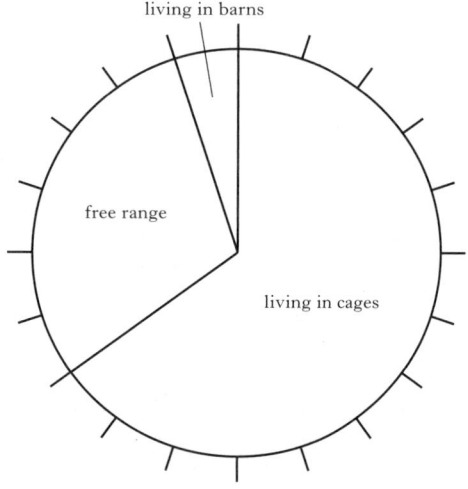

  (ii) 9 million / 9 000 000

(b) (i) 1400

  (ii) selective breeding

14. (a) (i)

NN	Nn
Nn	nn

  (ii) Parent 1   Polydactyly / Extra toes
    Parent 2   Normal / No extra toes /
          Non polydactyly
    F_1 Polydactyly / Extra toes

  (iii) discontinuous

(b) Fertilisation is random / Sample size too small / Number of offspring is too small

15. (a) (i) Continuous flow

  (ii) It is immobilised / It is attached onto beads **or** stationary surface

(b) (i) bacteria

  (ii) genes / chromosomal material / pieces of chromosome / DNA

(c) (i) sugar / maltose

  (ii) destroys / removes micro-organisms which might compete with yeast / prevents competition from unwanted micro-organisms

16. (a) (i) As lactose concentration decreases / lactic acid concentration increases

  (ii) Decrease in lactose greater than increase in lactic acid / More lactose is lost than lactic acid is produced

(b) 0.026

BIOLOGY CREDIT 2011

1. (a) Average Soil water content **or** description of that

   (b) (i) Named abiotic factor + source of error when measuring

   (ii) Several measurements of the abiotic factors taken / Average values were calculated

2. (a) (i) B
   C and D / D and C
   G

   (ii) (nitrifying) bacteria

   (b) 30

3. (a) (i) Diagram B

   (ii) Increased chance of pollen not achieving pollination / getting lost / not reaching destination
   To increase the chance of some of their pollen reaching other plants
   Low chance of pollination / less chance of pollination
   Ensure pollination takes place
   To increase change of pollination

   (b) More plants releasing pollen in May / grass, silver birch and oak releasing pollen / plants pollinating in April still releasing pollen / All species may be releasing pollen

   (c) (i)

	sexual	asexual
Variation exists amongst the offspring	✓	
Germination is not required		✓
Desirable characteristics are maintained		✓
Seeds are produced which can be dispersed	✓	

   (ii) Clone

4. (a) It increases up to 0·1 ppm. then it decreases.

   (b) To show that any result / change to the shoot growth was caused by the plant growth substance /
   To compare shoot growth with and without plant growth substance /
   To show that any change to the shoot growth was caused by the factor under investigation

5. (a) (i) Barley

   (ii) 162500

   (iii) Don't know total yield / tonnes / amount for wheat and barley / other crops

(b)
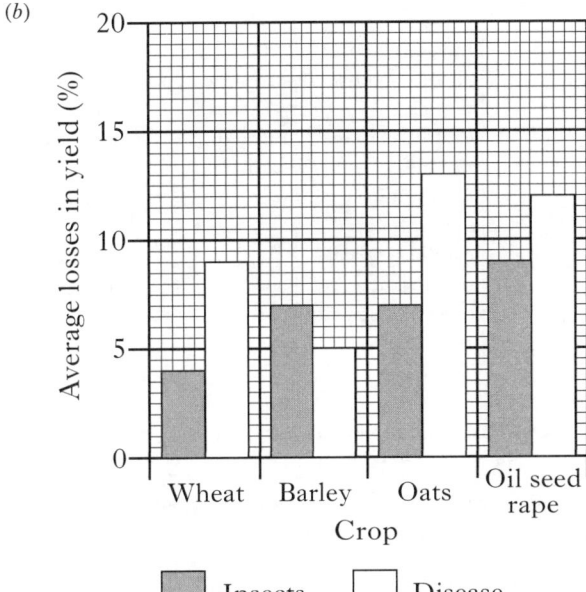

6. (a) 1. lichens
   2. dwarf mosses

   (b) Trees (or equivalent) / woodlands / forests have been cleared by humans / people

   (c) Average temperature during growing season / while plants are growing

   (d) 1. high winds
   2. wet conditions

   (e) Growing shoots are protected by surrounding vegetation

   (f) East

7. (a)
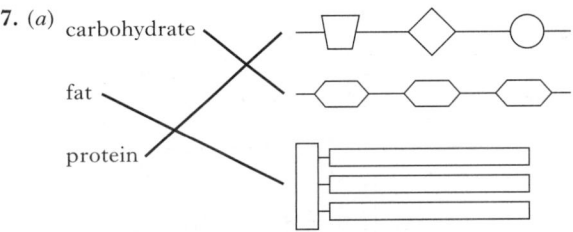

   (b)

B	F
D	E
C	

   (c) (i) Villus / villi

   (ii) A

   (iii) C

8. (a) (i)

Substance	oxygen	glucose	carbon dioxide
Importance	needed for respiration / to release energy / removal of waste	needed for respiration / energy source	removal of waste / needed for photosynthesis

(ii)

Location	lungs / alveoli / air sacs / cells / tissues / examples like muscles / placenta / Mesophyll / capillaries / cell membrane / red blood cells / stomata	villus / small intestine / cells / tissues / placenta / capillaries	Lungs / alveoli / air sacs / cells / examples of tissues / Mesophyll / placenta / capillaries / stomata

(b) Cell A

Cell has increased in volume / Cell is turgid / Cell is swollen / Cell vacuole has swollen

(c) Stage 2   Nuclear membrane disappears / breaks down **or**

Spindle forms **or**

Chromosomes / (pairs of) chromatids / they move to equator / middle of cell

Stage 4   Chromatids / they separate **or**

Chromatids / they are pulled apart **or**

Spindle fibres shorten

(d) So there is no loss of information / So they have the same information as parent cell

So they have a full set of information / genes / all genes passed on

9. (a) The other enzymes act on different substrates / substances

Enzymes are specific

(b) 4·3

(c) <u>Same</u> concentration of protein / <u>Same</u> concentration of agar / <u>Same</u> temperature /

<u>Same</u> volume **or** thickness of gel / <u>Same</u> concentration of enzyme / <u>Same</u> volume of enzyme (solution) / <u>Same</u> diameter **or** size of hole / same pH

10. (a)

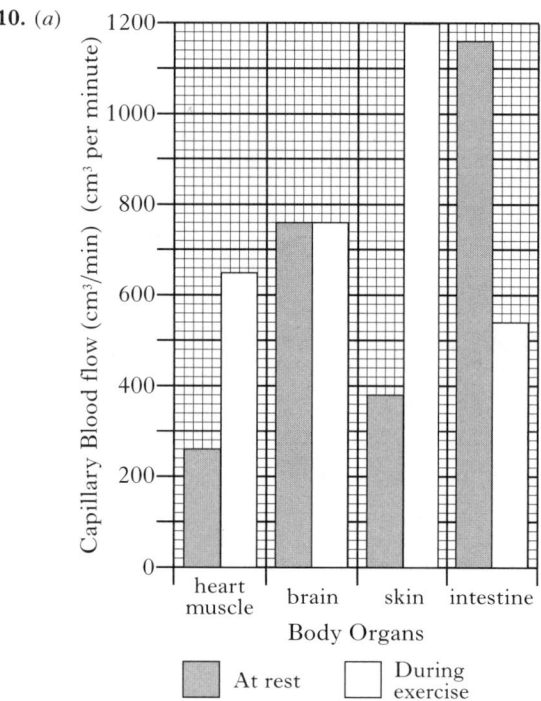

(b) 2 : 5

(c) To allow increased blood flow to other parts of body / skin / heart muscle / muscles

Other parts of body / skin / heart muscle / muscles need more blood

(d) Increase in blood flow (to skin during exercise)

11. (a) More heat lost to surroundings / air /

Less heat absorbed by test tube than by metal can

(b) Amount **or** mass **or** quantity of water / volume of water / same starting temperature of water / total burning of food / same distance between flame and container

12. (a) As the number of attempts increased, the performance / score improved until the 5th attempt

After that there was no further improvement / the performance remained the same

(b) Repeat with other people /

Repeat with same person after an interval

13. (a)

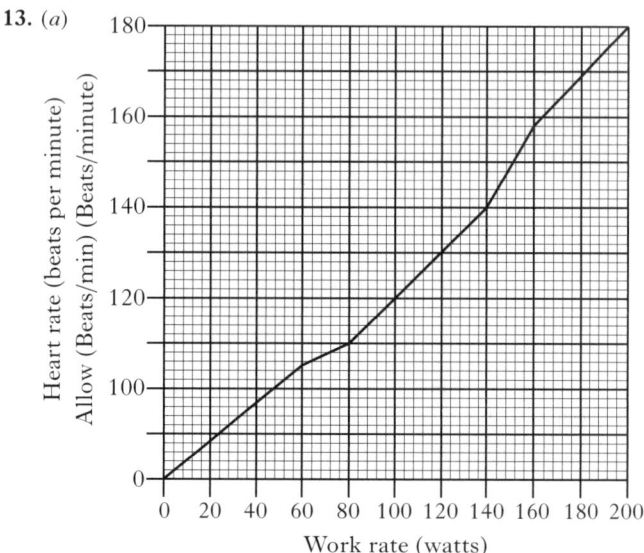

(b) 125

(c) (Heart and) lungs

14. (a) (i) Semi circular canals

(ii) They are at 90° to each other / They are at right angles to each other

(b) He would have better judgement of the distance (of the ball) / better depth of vision / better sense of distance / better perception of depth

(c) Sensory nerve cell → relay nerve cell → motor nerve cell → muscle

15. (a) Both (alleles) are the same / Both (alleles) are dominant **or** both are recessive

Only one form of allele / Identical alleles

Parents are either GG **or** gg

(b) genotype   Gg

phenotype   green

(c) (i) chromosomes / genes / chromatides

(ii) Radiation / atomic radiation / radioactivity / nuclear radiation / UV radiation / UV light / UV sunlight / X-rays / high temperatures / mustard gas / cochicine

(iii) 2162

(d) Selective breeding

16. (a) (i) Continuous flow (processing)

(ii) Enzymes / They can be reused / Enzymes do not need to be replaced

Product is easily separated /

No need to stop for cleaning / refilling reaction vessel

Cheaper

      (iii) Reduce the rate at which substract enters the vessel /
          slow down flow
          Reduce the rate at which the product leaves the vessel /
          Use more enzyme(s) / Increase the concentration of
          enzyme(s) /
          Decrease the concentration of the substrate / add less
          substrate
          Use smaller beads to increase S.A. of enzyme
          Put it through again / use a longer column of beads

(b) They are specific / No antibiotic can kill / act on all
    microbes / People may be allergic to some antibiotics /
    Bacteria can become resistant to antibiotics / New strains
    of bacteria appear

(c) (i) 7 am   3 pm   11 pm

    (ii) 63

17. (a) 0·1

   (b) (i) Remained the same / steady until 2003 / for the first
        3 years
        After that it increased

     (ii) 0·07

   (c) (i) carbon / carbon dioxide / phosphorus / phosphates /
        potassium / magnesium / calcium

     (ii) To kill <u>resistant</u> bacteria / fungal spores
        To kill endospores

Hey! I've done it

© 2011 SQA/Bright Red Publishing Ltd, All Rights Reserved
Published by Bright Red Publishing Ltd, 6 Stafford Street, Edinburgh, EH3 7AU
Tel: 0131 220 5804, Fax: 0131 220 6710, enquiries: sales@brightredpublishing.co.uk,
www.brightredpublishing.co.uk

Official SQA answers to 978-1-84948-159-5
2007-2011